BIRKHÄUSER

Igor Boiko

Discontinuous Control Systems

Frequency-Domain Analysis and Design

Birkhäuser
Boston • Basel • Berlin

Igor Boiko
Honeywell and University of Calgary
2500 University Dr. NW
Calgary, Alberta T2N 1N4
Canada
i.boiko@ieee.org

ISBN: 978-0-8176-4752-0   e-ISBN: 978-0-8176-4753-7
DOI: 10.1007/978-0-8176-4753-7

Library of Congress Control Number: 2008928863

© 2009 Birkhäuser Boston, a part of Springer Science+Business Media, LLC
All rights reserved. This work may not be translated or copied in whole or in part without the written permission of the publisher (Birkhäuser Boston, c/o Springer Science+Business Media, LLC, 233 Spring Street, New York, NY 10013, USA), except for brief excerpts in connection with reviews or scholarly analysis. Use in connection with any form of information storage and retrieval, electronic adaptation, computer software, or by similar or dissimilar methodology now known or hereafter developed is forbidden.
The use in this publication of trade names, trademarks, service marks and similar terms, even if they are not identified as such, is not to be taken as an expression of opinion as to whether or not they are subject to proprietary rights.

Printed on acid-free paper.

9 8 7 6 5 4 3 2 1

www.birkhauser.com

To my mother

Preface

Discontinuous control systems are the oldest type of control system and the most widespread type of nonlinear control system. The theory of discontinuous control, and the theory of relay feedback systems in particular, is usually considered a mature subject. However, many problems in discontinuous control theory still remain open. One problem involves the input-output properties of these systems, knowledge of which is extremely important to every application.

Two types of discontinuous control systems are studied in this book. The first is the so-called relay feedback system, which normally encompasses relay servomechanisms, various on-off controllers, sigma-delta modulators, relay feedback tests used for process dynamics identification, and controller tuning. Relay systems are often considered the main type of nonlinear system, which is evident by the enormous amount of house temperature control systems (that are usually implemented as on-off controllers) that exist. The theory of relay systems is an old subject. The problem of analysis of relay feedback systems was first considered by L. MacColl in 1945 [71]; the study was motivated by the development of relay servomechanisms of missile thrusters on the one hand and vibrational voltage regulators on the other. MacColl's analysis was based on an approximate approach close to the describing function method. Later, exact methods of analysis of relay feedback systems were developed, the most well-known of which is the Tsypkin locus [94]. The exact approach developed by Tsypkin, however, did not consider the servo aspect of relay feedback control. Its purpose was limited to finding periodic motions that may occur in a relay system in an autonomous mode or under external excitation. The servo problem in relay feedback control has not received due attention since. In recent years, a relatively small number of publications on relay feedback systems theory have appeared, in which only autonomous modes, and not servo modes, were considered.

The second type of discontinuous control system considered in this volume is the so-called sliding mode control system, which includes the conventional first-order sliding mode control system and the second-order sliding mode system. Sliding mode systems are a specific type of discontinuous control system. There are a number of references on sliding mode control (systems) devoted to this type of discontinuous control, the most well-known of which is by V.I. Utkin [97]. This subject has been an active research area during the past three decades. However, the present volume offers a different treatment of sliding modes than the traditional approach. The approach presented in this work allows accounting for the presence of so-called parasitic dynamics in control loops and uncovers mechanisms of chattering and non-ideal closed-loop performance in sliding mode control systems.

The purpose of this book is to present a new frequency-domain theory of discontinuous control systems in which the control systems are viewed and studied as servo systems. This theory involves a unified frequency-domain approach to both analysis of possible self-excited periodic motions and analysis of input-output properties of discontinuous control systems. The servo aspect of control is very important and was underestimated in the past. Knowledge of input-output properties is as important as knowledge of autonomous behavior in discontinuous control systems (self-excited oscillations). In fact, these two aspects complement each other. For example, in on-off house temperature control system design, it is equally important to know both the frequency of relay switching and how the average indoor temperature might change in response to the outdoor temperature. The latter problem can be solved only if the servo aspect of the system is considered.

The core approach in this present book is the frequency-domain method called the locus of a perturbed relay system (LPRS). This method offers an exact analysis of both oscillatory properties and servo properties of relay feedback systems. The method is analytical and very convenient for design applications, which is illustrated by the numerous examples provided. This approach is exact, which allows for overcoming the drawback of the well-known describing function method. Further, overcoming the constraint of the filtering hypothesis of the describing function method allows for extending the proposed theory to analysis of motion in sliding mode control systems, where the sliding mode itself is now considered as oscillations of either finite or infinite frequency. The analysis provided, however, is not merely another confirmation of available results in sliding mode control theory. The proposed approach offers a more precise treatment of sliding mode control systems than does classic sliding mode control theory. Thus, the proposed approach introduces the theory of *real* sliding mode control versus *ideal* classic sliding mode control.

This book is primarily a research monograph, as it is devoted only to frequency-domain theory of discontinuous control systems, and the theory presented in the book is novel. However, it also has many features of a textbook, as the theory presented covers a relatively large classic nonlinear control area. This theory is illustrated by a number of application examples from different areas of control engineering and is accessible to students with a background in linear control. The included MATLAB code can also make understanding of the presented theory easy. The book, therefore, can be used by researchers, practitioners, and undergraduate and graduate students.

The material is organized into two parts and appendices. Part I is devoted to the theory of the locus of a perturbed relay system (LPRS) method, and Part II presents applications of the LPRS method. Part I comprises five chapters. Chapter 1 poses the problem and outlines the scope and method of analysis, which is presented in detail in the following chapter. In Chapter 3, the results obtained in Chapter 2 for slow inputs are extended to the analysis of the system response to relatively fast input signals. Chapter 4 presents frequency-domain theory of sliding mode control systems. The sliding mode analysis follows the methodology developed in the preceding chapters. Chapter 5 is devoted to an emerging area of sliding mode control — second-order sliding mode control systems analysis.

Part II gives a number of applications of LPRS theory. These applications include the electro-pneumatic servomechanism (Chapter 6), the relay feedback test and its application to process dynamics identification and controller tuning (Chapter 7), the sliding mode differentiator and dynamic compensator (Chapter 8), and the sliding mode observer (Chapter 9). Some of these applications have been in engineering practice for many years, and other applications are relatively new. In general, the material of each chapter in the second part of the book is independent and only makes general references to topics presented in the first part. The Appendix contains the derivations of the LPRS, the proofs of the theorems, and the MATLAB code used in the text for the LPRS computations.

I wish to acknowledge the many colleagues who have contributed to the development of this book. I am especially grateful to Prof. L. Fridman for the many discussions resulting in the refinement of the theory presented, to Dr. A. Pisano for his review of the manuscript, to Prof. V.I. Utkin and Prof. K. Furuta for their encouragement and support of my work on this project, to the late Prof. V. Kainov who years ago introduced me to the world of discontinuous control systems, to Prof. N.V. Faldin who influenced my research in many ways, to Prof. D. Atherton and Prof. Y. Shtessel for their valuable comments and discussions of the theory currently presented in the book, and to Mr. E. Tamayo for providing the opportunity and help in implementation of the presented theory in a loop tuner. I thank Alex and Michael Boiko for their help with the LaTeX typesetting and artwork. I also express gratitude to my family for their patience and support; without them this undertaking would not have been possible.

Calgary, Canada *Igor Boiko*
October 2008

Contents

Preface VII

Part I The locus of a perturbed relay system theory

1 The servo problem in discontinuous control systems 3
   1.1 Introduction 3
   1.2 Fundamentals of frequency-domain analysis of periodic motions in nonlinear systems............................... 5
   1.3 Relay servo systems...................................... 10
   1.4 Symmetric oscillations in relay servo systems: DF analysis 12
   1.5 Asymmetric oscillations in relay servo systems: DF analysis ... 14
   1.6 Slow signal propagation through a relay servo system 16
   1.7 Conclusions.. 17

2 The locus of a perturbed relay system (LPRS) theory 19
   2.1 Introduction to the LPRS 19
   2.2 Computing the LPRS for a non-integrating plant 21
      2.2.1 Matrix state-space description approach 21
      2.2.2 Partial fraction expansion technique 23
      2.2.3 Transfer function description approach 24
      2.2.4 Orbital stability of relay systems 26
   2.3 Computing the LPRS for an integrating plant 26
      2.3.1 Matrix state-space description approach 26
      2.3.2 Transfer function description approach 29
      2.3.3 Orbital stability of relay systems 30
   2.4 Computing the LPRS for a plant with a time delay 31
      2.4.1 Matrix state-space description approach 31
      2.4.2 Orbital asymptotic stability 32
   2.5 LPRS of first-order dynamics 33
   2.6 LPRS of second-order dynamics 35

	2.7	LPRS of first-order plus dead-time dynamics	38
	2.8	Some properties of the LPRS	41
	2.9	LPRS of nonlinear plants	43
		2.9.1 Additivity property	43
		2.9.2 The LPRS extended definition and open-loop LPRS computing	46
	2.10	Application of periodic signal mapping to computing the LPRS of some special nonlinear plants	48
	2.11	Comparison of the LPRS with other methods of analysis of relay systems	52
	2.12	An example of analysis of oscillations and transfer properties	53
	2.13	Conclusions	54
3	**Input-output analysis of relay servo systems**	57
	3.1	Slow and fast signal propagation through a relay servo system	57
	3.2	Methodology of input-output analysis	63
	3.3	Example of forced motions analysis with the use of the LPRS	63
	3.4	Conclusions	65
4	**Analysis of sliding modes in the frequency domain**	67
	4.1	Introduction to sliding mode control	67
	4.2	Representation of a sliding mode system via the equivalent relay system	69
	4.3	Analysis of motions in the equivalent relay system	73
	4.4	The chattering phenomenon and its LPRS analysis	77
	4.5	Reduced-order and non–reduced-order models of averaged motions in a sliding mode system and input-output analysis	85
	4.6	On fractal dynamics in sliding-mode control	88
	4.7	Examples of chattering and disturbance attenuation analysis	95
	4.8	Conclusions	101
5	**Performance analysis of second-order SM control algorithms**	103
	5.1	Introduction	103
	5.2	Sub-optimal algorithm	104
	5.3	Describing function analysis of chattering	105
	5.4	Exact frequency-domain analysis of chattering	106
	5.5	Describing function analysis of external signal propagation	108
	5.6	Exact frequency-domain analysis of external signal propagation	112
	5.7	Example of the analysis of sub-optimal algorithm performance	117
	5.8	Conclusions	122

Part II Applications of the locus of a perturbed relay system

6 Relay pneumatic servomechanism design 125
  6.1 Relay pneumatic servomechanism dynamics
      and characteristics 125
  6.2 LPRS analysis of uncompensated relay electro-pneumatic
      servomechanism 127
  6.3 Compensator design in the relay electro-pneumatic
      servomechanism 128
  6.4 Examples of compensator design in the relay
      electro-pneumatic servomechanism 132
  6.5 Compensator design in the relay electro-pneumatic
      servomechanism with the use of the LPRS
      of a nonlinear plant 135
  6.6 Conclusions 138

7 Relay feedback test identification and autotuning 139
  7.1 The relay feedback test 139
  7.2 The LPRS and asymmetric relay feedback test 140
  7.3 Methodology of identification of the first-order plus
      dead-time process 141
  7.4 Analysis of potential sources of inaccuracy 143
  7.5 Performance analysis of the identification algorithm 145
  7.6 Tuning algorithm 147
  7.7 Conclusions 151

**8 Performance analysis of the sliding mode–based analog
   differentiator and dynamical compensator** 153
  8.1 Transfer function "inversion" via sliding mode 153
  8.2 Analysis of SM differentiator dynamics 154
  8.3 Temperature sensor dynamics compensation
      via SM application 157
  8.4 Analysis of the sliding mode compensator 160
  8.5 An example of compensator design 162
  8.6 Conclusions 165

9 Analysis of sliding mode observers 167
  9.1 The SM observer as a relay servo system 167
  9.2 SM observer performance analysis and characteristics 170
  9.3 Example of SM observer performance analysis 172
  9.4 Conclusions 175

10 Appendix 177
10.1 The LPRS derivation for a non-integrating linear part 177
10.2 Orbital stability of a system with a non-integrating linear part 181
10.3 The LPRS derivation for an integrating linear part 183
10.4 Orbital stability of a system with an integrating linear part ... 191
10.5 The LPRS derivation for a linear part with time delay 194
10.6 MATLAB code for LPRS computing 198

References 205

Index 211

Part I

The locus of a perturbed relay system theory

1
The servo problem in discontinuous control systems

1.1 Introduction

Discontinuous control systems, and relay feedback systems in particular, are one of the most important types of nonlinear systems. The term "relay" comes from electrical applications where on-off control has long been used. To describe the nonlinear phenomenon typical of electrical relays, the nonlinear function also received the name "relay" comprising a number of discontinuous nonlinearities. Thus, when applied to any type of control system, the term "relay" is now associated not with applications but with the kind of nonlinearities that are found in the system models. However, what we traditionally call the "relay system" cannot always be described by the relay system model. For example, the vibrational voltage regulator is not a relay system in the full sense, as the charge and discharge time constants are different and the regulator is better described as a variable structure system (which is, however, a discontinuous control system).

Applications of the discontinuous feedback principle have evolved from vibrational voltage regulators and missile thruster servomechanisms of the 1940s to numerous on-off process parameter closed-loop control systems, sigma-delta modulators, process identification and automatic tuning of proportional-integral-derivative (PID) controller techniques, and DC motor, hydraulic and pneumatic servo systems, to name a few. The enormous number of residential temperature control systems available throughout the world illustrates how popular discontinuous control systems are. A number of industrial examples of relay systems were given in the classic book on relay systems [94]. Furthermore, many existing sliding mode algorithms can be considered and analyzed via the relay control principle. Perhaps, some aspects of hybrid systems can also be analyzed via application of discontinuous control system theory. In fact, discontinuous control systems are probably the most conventional type of control system in history.

Discontinuous control systems provide many advantages over linear systems: simplicity of design, cheaper components that were known as early as the 1930s [55], and the ability to adapt the open-loop gain in the relay feedback system in response to changing parameters [8, 50, 71]. As a rule, they also provide a higher open-loop gain and a higher performance [56] than linear systems. In some applications, smoothing of the Coulomb friction and of other plant nonlinearities can also be achieved.

Discontinuous control systems theory has been the subject of attention since the 1940s from the worldwide research community. Traditionally, the scope of research was composed of the following problems: existence and parameters of periodic motions, stability of limit cycles, and input-output problems (set point tracking or external signal propagation through the system), which includes the disturbance attenuation problem. The theoretical development was motivated by the design of missile thruster servomechanisms in Germany [17] and in the United States [71, 72, 104] as well as vibrational voltage regulators [51]. Later, a number of publications, some of which became classic works in the area, appeared. Most of these were concerned with the solution of the periodic problem. A semigrahpical solution [43], a matrix method [28], a z-transform–based solution [57], frequency-domain methods [54, 94[1]] and [9] (which can be considered a generalization of method [94]), a state-space–based technique [34], and a finite-difference operator method [81] for the solution of problem of existence and local stability of periodic motions were proposed in the late 1950s to the mid-1960s. In [79] a systematic overview of applications of Poincaré maps to the analysis of periodic motions in relay systems was given. A rigorous solution of the problem of existence and local stability of limit cycles was presented in [2] and [100] for symmetric and asymmetric limit cycles, respectively. The problem of global stability of limit cycles was considered in [53]. In [81, 94] and [84], the input-output problem was analyzed. Relay feedback systems were also studied with the use of more general approximate methods such as the describing function (DF) method [71] (where the input-output problem in relay systems was probably considered for the first time; the authors used the results obtained in [58]), and [8, 50, 56, 87]. In the recent monographs [103] and [107], extensive coverage of the theory of relay feedback test–based identification is given.

Another study of oscillations in the relay and variable structure systems was motivated by the chattering problem in sliding mode control [97, 105]. This problem is closely related to the original problem, and the results are applicable to both variable structure systems and relay systems ([4, 16, 26, 46, 47, 98, 99]). A number of techniques of process parameter identification and PID controller tuning based on the relay feedback test are given in [3, 59–61, 70, 73–75, 83]. Periodic motions in sigma-delta modulators were studied in [41] and [78].

The method presented in this book is based on the frequency-domain characteristic of the discontinuous control system that is similar in some ways to the frequency response of a linear system (Nyquist plot) and the Tsypkin

---

[1] The first edition of this book was published in 1955.

locus [94]. Therefore, the presented approach has some resemblance to the DF method and Tsypkin's method. Moreover, from the methodological point of view, the presented theory in some ways replicates the DF method. However, the presented method offers a few advantages over both methods mentioned above: better accuracy than that of the DF method (eliminating the necessity of the filtering hypothesis) and a better solution of the input-output problem compared with that of the Tsypkin locus, which can only furnish a solution of the periodic problem.[2]

All discontinuous control systems have a nonlinear element in the control loop. Usually the nonlinearity is associated with the controller. In most cases the discontinuous nonlinearity is an ideal (sign function) or hysteretic relay. However, it can be a more complex nonlinearity, examples of which are considered in the chapters devoted to sliding mode control and second-order sliding mode control.

A mode that may occur in a discontinuous control system is a self-excited (non-vanishing) oscillation, which is also referred to as a periodic motion. If the system does not have asymmetric nonlinearities, this periodic motion is symmetric in the autonomous mode (no external input applied). However, if an external input (disturbance) is applied to the system that has a periodic motion, the self-excited oscillations become biased or asymmetric. The key approach to the analysis of discontinuous control systems is a method of analysis of self-excited oscillations (symmetric and asymmetric). The following section gives a general methodology of such an analysis based on frequency-domain concepts.

1.2 Fundamentals of frequency-domain analysis of periodic motions in nonlinear systems

The problem of finding possible periodic motions in nonlinear dynamical systems is a fundamental problem in both mathematics and control theory. It can be traced back to the works of Poincaré who introduced a widely used now geometric interpretation of this problem as that of finding a closed orbit in the state space.

Poincaré's approach involves finding a certain map in the state space, which is now called the Poincaré map. Yet Poincaré maps for most nonlinear systems can hardly be obtained in an analytical form. Analytical expressions have been obtained only for low-order systems or higher-order systems containing simple types of nonlinear functions. An example of the latter is the relay feedback system for which the Poincaré map is obtained in an explicit form. A few formulations of Poincaré maps for relay feedback systems are presented in [2, 20, 54, 94, 100]. For more complex types of nonlinearities, only an

---
[2] Tsypkin [94] also considers the input-output problem in his book; he proposes that the solution of the input-output problem should be done via analysis of asymmetric oscillations.

approximate solution can be found as a rule. A number of different approaches to the problem of finding a periodic solution have been proposed. The most general approach was given in [64] in which the solution of the periodic problem was found in the form of a Fourier series. Despite the clear underlying idea of this method, its application is computationally cumbersome, as the original problem is transformed into the problem of solving the system of a high number of nonlinear algebraic equations, which is often impractical. A modification of this classic approach was given in [96]. However, the computational efficiency of the proposed approach rapidly decreases with required accuracy. Among other methods are Volterra series used to solve both the periodic problem in autonomous systems [32] and forced oscillations [102]. Yet the Volterra series approach can only handle weak (mild) nonlinearities. A combination of the perturbation method and the harmonic balance method was proposed in [29]. But this method also can handle only mild nonlinearities. A singular perturbation method was used for analysis of the periodic problem in relay systems in [47].

Commonly in engineering practice, only the first-harmonic approximation is used, which is known as the describing function (DF) method. The main concepts of this method were developed in the 1940s and 1950s [36, 51, 52, 63, 71, 95]. Comprehensive coverage of this method is given in [8, 50]. The DF method provides a simple and efficient, but not exact, solution of the periodic problem. Its applications are limited to systems that can be separated into nonlinear and linear parts interconnected into a closed loop (Fig. 1.1) with the nonlinear part being a static nonlinearity (with possible hysteresis) and the linear part given by linear time-invariant differential equations. Moreover, the linear part must be a low-pass filter to filter out higher harmonics, so that the input to the nonlinearity could approximately be considered a sinusoid.

Let us consider the frequency-domain approach to analysis of self-excited oscillations. Consider the following nonlinear system,

$$\dot{\mathbf{x}} = \mathbf{A}\mathbf{x} + \mathbf{B}u$$
$$y = \mathbf{C}\mathbf{x} \tag{1.1}$$

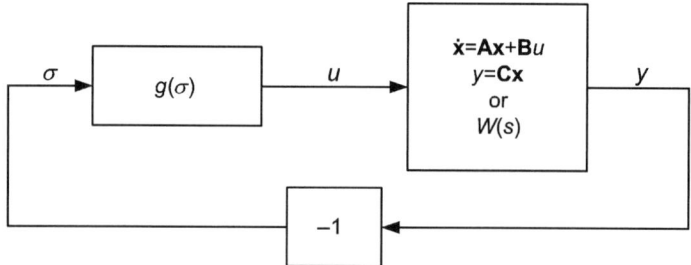

Fig. 1.1. Block diagram of the system for DF analysis

1.2 Fundamentals of frequency-domain analysis of periodic motions

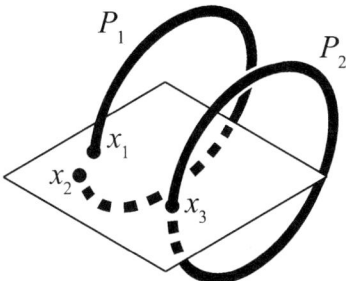

Fig. 1.2. Poincaré mapping

$$u = f(\sigma), \qquad (1.2)$$

$$\sigma = -y, \qquad (1.3)$$

where $\mathbf{A} \in R^{n \times n}$, $\mathbf{B} \in R^{n \times 1}$, and $\mathbf{C} \in R^{1 \times n}$ are matrices, $\mathbf{x} \in R^n$, $y \in R^1$, $\sigma \in R^1$, $f \in R^1$, $f(\sigma)$ is a single-valued symmetric monotone nonlinear function: $f(-\sigma) = -f(\sigma)$ (it will be later extended to a hysteretic function). Formula (1.1) provides a description of the linear part, formula (1.2) denotes the nonlinear part of the system, and formula (1.3) gives the condition of closing the system loop. We shall also use the linear part description in the form of the transfer function $W(s)$, which can be obtained from formulas (1.1) as follows: $W(s) = \mathbf{C}(\mathbf{I}s - \mathbf{A})^{-1}\mathbf{B}$. Let us also assume that the linear part is strictly proper, i.e., the relative degree of $W(s)$ is one or higher.

The problem of finding a periodic solution of system (1.1), (1.2), and (1.3) is usually formulated as the problem of finding a fixed point $\mathbf{x}^*$ of Poincaré map, which will be a solution to the equation: $\mathbf{x} = P(\mathbf{x})$, where $P(\mathbf{x})$ is the mapping. Figure 1.2 provides two examples of Poincaré mapping. P_1 maps point $\mathbf{x}_1$ into point $\mathbf{x}_2$, and P_2 maps point $\mathbf{x}_3$ into itself. Therefore, $\mathbf{x}_3 = \mathbf{x}^*$. In the time domain, the closed orbit that starts from the point $\mathbf{x}^*$ (denote it $\mathbf{x}^*(t)$ for brevity) will be a periodic vector signal, and $y^*(t) = \mathbf{C}\mathbf{x}^*(t)$ will be a periodic scalar signal.

Now consider the open-loop system (1.1), (1.2), which can be obtained by ignoring equality (1.3). Let $\sigma(t)$ be a periodic symmetric signal of period T with zero mean, continuously differentiable with respect to t, so that $\sigma(t + T/2) = -\sigma(t)$, $\sigma \in R$, $t \in D \subset R$, $D := t \in [0; T]$. Consider the following lemma, which will be instrumental below.

Lemma 1.1. *If $\sigma(t)$ satisfies the conditions of symmetric periodicity and unimodality and $f(\sigma)$ is a symmetric monotone nonlinear function, then $\mathbf{x}(t)$ will be a periodic vector signal, and $y(t)$ will be a periodic scalar signal.*

Proof. $u(t)$ is a periodic signal because it is a function of $f(\sigma)$: $u(t+T) = f(\sigma(t+T)) = f(\sigma(t)) = u(t)$. $y(t)$ as well as $\mathbf{x}(t)$ are periodic because they are responses to the periodic input $u(t)$, which follows from linear system theory. ∎

Definition 1.2. *Let us call the mapping of space $R \times D$ into itself, where $\sigma(t)$ satisfies the conditions of symmetric periodicity and unimodality, the periodic signal mapping (PSM)*

$$y(t) = g(f(\sigma(t))), \tag{1.4}$$

which is a chain-rule application of two mappings: g for the linear part and f for the nonlinear part.

Definition 1.3. *Let us call the periodic function $\sigma^*(t)$, which provides the solution to the equation $-\sigma(t) = g(f(\sigma(t)))$, $t \in D$, the fixed point of the periodic signal mapping (the minus sign is attributed to the negative feedback in the closed-loop system).*

The following theorem relates periodic solutions in the state-space and time domains.

Theorem 1.4. *If $\sigma^*(t)$ is a fixed point of the periodic signal mapping $y(t) = g(f(\sigma(t)))$ in the open-loop nonlinear system (1.1), (1.2) where $f(\sigma)$ is a single-valued symmetric monotone nonlinear function, then $\mathbf{x}^*(t) = h(f(\sigma^*(t)))$ will be a fixed point of the Poincaré mapping in the closed-loop system (1.1), (1.2), (1.3), where h is mapping $u(t) \to \mathbf{x}(t)$.*

Proof. Because $\sigma^*(t)$ is a fixed point of the periodic signal mapping (1.4), $\sigma^*(t)$ provides a solution to system (1.1), (1.2), (1.3). This is a periodic solution, as the property of periodicity is "embedded" in the type of function $\sigma(t)$. According to Lemma 1.1, $\mathbf{x}^*(t)$ will be a periodic vector function, which means that in the state space, the trajectory that starts from $\mathbf{x}^*$ is a closed orbit, i.e., satisfying the condition of being a fixed point of the Poincaré map. ∎

Let us analyze how the PSM can be implemented in practice. Transfer the analyzed problem including the defined mappings into the spectral domain, which can be provided via taking the Fourier series of the signals in the system. Represent the periodic signal $\sigma(t)$ in the form of the Fourier series as follows,

$$\sigma(t) = 2 \sum_{k=1}^{\infty} |q_{\sigma k}| \cos\left((2k-1)\omega t + \arg q_{\sigma k}\right) \tag{1.5}$$

where $\omega = 2\pi/T$, and all even harmonics are zeros due to $\sigma(t)$ being symmetric

$$q_{\sigma k} = \frac{1}{T} \int_{-T/2}^{T/2} \sigma(t) \left[\cos(2k-1)\omega t - j\sin(2k-1)\omega t\right] dt. \tag{1.6}$$

Therefore, any $\sigma(t)$ can be characterized by the matrix $\mathbf{Q}_\sigma$ as follows: $\mathbf{Q}_\sigma = [q_{\sigma 1}\ q_{\sigma 3}\ q_{\sigma 5} \ldots]$ subject to the frequency ω being given. We shall call the matrix $\mathbf{Q}_\sigma$ the spectrum of the signal $\sigma(t)$. The number of elements of $\mathbf{Q}_\sigma$ can be set arbitrarily large, but it makes sense to limit the number of elements

1.2 Fundamentals of frequency-domain analysis of periodic motions

to some reasonable value, so that any further increase of this number will not provide any noticeable increase in the accuracy of the $\sigma(t)$ representation. Let us represent other periodic signals u and y in the system by their spectra: $\mathbf{Q}_u$, $\mathbf{Q}_y$. Then the mapping $\mathbf{Q}_\sigma \to \mathbf{Q}_u$ can be given by the transformation into the time domain as per (1.5), nonlinear mapping in the time domain as per (1.3), and transformation into the spectral domain through the following formulas:

$$q_{uk} = \frac{1}{T} \int_{-T/2}^{T/2} u(t) \left[\cos(2k-1)\omega t - j \sin(2k-1)\omega t \right] dt \qquad (1.7)$$

$$\mathbf{Q}_u = [q_{u1} \; q_{u3} \; q_{u5} \ldots]. \qquad (1.8)$$

The mapping $\mathbf{Q}_u \to \mathbf{Q}_y$ can be given by the following element-by-element multiplication of the two matrices, which directly follows from linear systems theory:

$$\mathbf{Q}_y = \mathbf{Q}_u \bullet \mathbf{S} \qquad (1.9)$$

where

$$\mathbf{Q}_y = [q_{y1} \; q_{y3} \; q_{y5} \ldots] \qquad (1.10)$$

$$\mathbf{S} = [W(j\omega) \; W(j3\omega) \; W(j5\omega) \ldots]. \qquad (1.11)$$

According to Theorem 1.4, the periodic solution of (1.1), (1.2), (1.3) can be found as a fixed point of the PSM. We have defined the PSM as the chain application of two mappings: the first of which is provided by (1.5), (1.2), and (1.7), and the second is given by (1.9). To find the fixed point, we have to design a contraction mapping, so that iterates from the initial point will converge to the solution.

Assume now that the linear part is an ideal low-pass filter. This assumption enables one to rewrite (1.9) as

$$\mathbf{Q}_y = q_{u1} W(j\omega) \qquad (1.12)$$

and (1.5), (1.2), (1.7) as just

$$q_{u1} = \frac{1}{T} \int_{-T/2}^{T/2} u(t) \left[\cos \omega t - j \sin \omega t \right] dt \qquad (1.13)$$

with $\sigma(t) = a_\sigma \cos \omega t$. The ratio $q_{u1}/q_{\sigma 1} = 0.5 a_\sigma$ gives the describing function (DF) of the nonlinearity. As a result, considering (1.3) we can write the harmonic balance equation for the DF approximated system as follows:

$$N(a_\sigma) W(j\Omega) = -1 \qquad (1.14)$$

where $N(a_\sigma)$ is the describing function of the nonlinearity, a_σ is the amplitude of $\sigma(t)$, and Ω is the frequency of the periodic motion:

$$N(a_\sigma) = \frac{q_{u1}}{q_{\sigma 1}} = \frac{2}{Ta_\sigma} \int_{-T/2}^{T/2} u(t)\left[\cos\omega t - j\sin\omega t\right] dt.$$

Equation (1.14) is a well-known formula [8, 50] that is widely used for analysis of periodic motions. Therefore, the DF analysis can be viewed as a fixed point of the PSM considered above, subject to the assumption about the low-pass filtering property of the linear part.

Equation (1.14) provides a basis for analysis of possible periodic motions in a nonlinear system via application of the DF method. Usually this equation is presented in the following form:

$$W(j\Omega) = -\frac{1}{N(a_\sigma)} \qquad (1.15)$$

which is convenient for graphical interpretation, as the unknown variables are separated into the left-hand and the right-hand parts of the equation.

1.3 Relay servo systems

Usually, control systems are categorized into two groups depending on the control problem they are supposed to solve. These two categories are *stabilization systems* and *servo systems*. Stabilization systems carry out the task of stabilization of certain process variables. The main challenge for such systems is producing a compensating action for the disturbances that inevitably occur. The servo systems implement the task of tracking certain input signals, so that the system output follows the input signal as precisely as possible. Normally, the input to the system is unknown, but some *a priori* information about the input signal such as frequency and amplitude range may be available. The typical examples of stabilization systems are temperature, pressure, level, flow control systems; and examples of servo systems are actuators, airplane autopilots, and temperature control systems, which are supposed to ensure the temperature change in time according to a given profile.

We can see from the above examples that the *autonomous* mode, when no external signals are applied to the system, does not normally occur. What is common for both these types of control systems is the existence of external signals that affect the functioning of the system. In the first case, this is a disturbance to which the system is supposed to respond in such a way as to provide its compensation. In the second case, the system is supposed to respond to the input to bring the output in alignment with this external input. In both cases, the problem analyzing the effect of external signals on the system characteristics is the most important part of system performance analysis.

1.3 Relay servo systems

One could argue that for a stabilization system, this analysis is of secondary importance. The example that may be given is control loop tuning from the process step response [109]. Indeed, the disturbance effect is not analyzed in this case. However, the logic of this tuning is as follows: to apply the step test to identify some essential performance characteristics of the loop and to tune this loop in such a way that the best disturbance rejection is obtained. Set point change is not typical of stabilization systems, and the step response is just a certain characteristic of a servo system, but the principal concern when performing this test and the subsequent tuning, is the system response to a potential disturbance. The explanation of such methodology lies in the fact that it is much easier to generate the set point change than the disturbance change. The disturbance for a temperature control system may be, for instance, the ambient temperature, and it is practically impossible in most cases to create the ambient temperature change to be able to measure the system response.

This issue is often forgotten, creating the illusion that the autonomous mode is enough for analysis of stabilization system performance. This is especially true with respect to the relay feedback systems and sliding mode (SM) systems, where the input-output problem did not receive adequate attention. A few examples where this problem was considered are given in the introduction. A vast majority of the publications on relay system theory are devoted to the analysis of the parameters of the oscillations and their stability. In SM theory, only chattering as a phenomenon typical of real sliding received some attention as a subject of research.

By the *servo problem* we mean the analysis of the system response to an external signal: either this external signal is the disturbance, which the system is supposed to reject, or it is the input, which the system is supposed to track. From this definition, one can see that this problem applies to both types of systems considered above. Moreover, if we consider a model of the system, the difference between those two signals is only in the point of application, and from the point of the methodology of analysis they are no different. Because we deal with models in this book, we can consider only one signal applied to the system and consider it as being either a disturbance or a reference input signal, depending on the system task. Naturally, the servo problem cannot be solved without the autonomous mode analysis having been carried out first. The servo problem is, therefore, an extension of the analysis of the system in an autonomous mode. It includes the autonomous mode analysis and provides, therefore, a more complex type of analysis.

We call the system described by the following equations the *relay servo system* — emphasizing the fact that an external input is applied to this system, and the system response to this input is one of the subjects of analysis,

$$\dot{\mathbf{x}} = \mathbf{A}\mathbf{x} + \mathbf{B}u$$
$$y = \mathbf{C}\mathbf{x}$$
(1.16)

12   1 The servo problem in discontinuous control systems

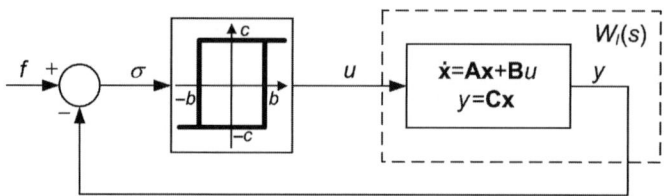

Fig. 1.3. Relay servo system

$$u = \begin{cases} +c \text{ if } \sigma = f_0 - y \geq b \text{ or } \sigma > -b, u(t-0) = c \\ -c \text{ if } \sigma = f_0 - y \leq -b \text{ or } \sigma < b, u(t-0) = -c, \end{cases}$$

where $\mathbf{A} \in R^{n \times n}, \mathbf{B} \in R^{n \times 1}, \mathbf{C} \in R^{1 \times n}$ are matrices, and $u(t-0)$ is the control value at the time immediately preceding the current time.

We represent the relay servo system as a block diagram (Fig. 1.3). In Fig. 1.3, f is a cumulative input (disturbance) to the system transposed to the relay input, u is the control, y is the output, σ is the error signal, c is the amplitude of the relay, $2b$ is the hysteresis value of the relay function $u = u(\sigma)$, and $W_l(s)$ is the transfer function of the linear part (of the plant in the simplest case), which can be obtained from the matrix-vector description (1.16) as $W_l(s) = \mathbf{C}(\mathbf{I}s - \mathbf{A})^{-1}\mathbf{B}$.

1.4 Symmetric oscillations in relay servo systems: DF analysis

The describing function (DF) method provides a simple and often fairly precise approach to the problems of analysis of periodic motions and input-output analysis (within the framework of the assumption about a sinusoidal input to the relay). The exact analysis considered below has many common features with the DF analysis. For that reason, a review of the DF analysis of system Fig. 1.3 is beneficial for understanding the concepts of the method described in the subsequent chapters of this book. Consider the analysis of possible periodic motions in the relay servo system (Fig. 1.3).

Consider the autonomous mode. Assume that the input to the system is identically equal to zero ($f(t) \equiv 0$). Then we can assume that a symmetric periodic process of unknown frequency Ω_p and amplitude a_p of the input to the relay occurs in the system. Finding the values of the frequency and amplitude is the main objective of this analysis.

In accordance with DF method concepts, we assume that the input to the nonlinearity is a harmonic signal, and the so-called describing function of the nonlinearity can be written as a function of the amplitude and frequency as follows:

$$N(a,\omega) = \frac{\omega}{\pi a} \int_0^{2\pi/\omega} u(t) \sin \omega t \, dt + j \frac{\omega}{\pi a} \int_0^{2\pi/\omega} u(t) \cos \omega t \, dt. \quad (1.17)$$

1.4 Symmetric oscillations in relay servo systems: DF analysis

The DF given by formula (1.17) is essentially a complex gain of the transformation of the harmonic input by the nonlinearity into the control signal with respect to the first harmonic in the control signal. For the hysteretic relay nonlinearity, the formula of the DF can be obtained analytically. It is given as follows [8]:

$$N(a) = \frac{4c}{\pi a}\sqrt{1 - \left(\frac{b}{a}\right)^2} - j\frac{4cb}{\pi a^2}, (a \geq b). \tag{1.18}$$

For the hysteretic relay, the DF is a function of the amplitude only and does not depend on frequency.

The periodic solution in the relay feedback system can be found from the equation of *harmonic balance* [8],

$$W_l(j\Omega_p) = -\frac{1}{N(a_p)}, \tag{1.19}$$

which is a complex equation with two unknown values: frequency Ω_p and amplitude a_p. Equation (1.19) has a convenient graphical interpretation. Note that the value on the left-hand side of the equation is considered a function of the frequency, and in fact is the Nyquist locus of the linear part of the system. The value on the right-hand side is the negative reciprocal of the DF of the hysteretic relay. Obtain the negative reciprocal of the DF from formula (1.18):

$$-N^{-1}(a) = -\frac{\pi a}{4c}\sqrt{1 - \left(\frac{b}{a}\right)^2} - j\frac{\pi b}{4c}, (a \geq b). \tag{1.20}$$

We can see from (1.20) that the imaginary part does not depend on the amplitude a, and the plot of $-N^{-1}(a)$ on the complex plane is a horizontal line (Fig. 1.4), which lies on the left half-plane. The point corresponding to zero amplitude is located on the imaginary axis, and the real part of $-N^{-1}(a)$ tends to minus infinity as the amplitude grows.

The periodic solution of the equations of the relay servo system corresponds with the point of intersection of the Nyquist plot of the linear part (being a function of the frequency) and of the negative reciprocal of the DF of the hysteretic relay (being a function of the amplitude), given by formula (1.20), on the complex plane. This periodic solution is approximate, a result of the approximate nature of the DF method itself, which is based upon the assumption about the harmonic shape of the input signal to the relay. However, if the linear part of the system has the property of the low-pass filter, so that the higher harmonics of the control signal are attenuated to a higher degree, the DF method may give a relatively precise result in terms of the found values of the frequency and the amplitude. A comparison with the exact solution is given in Chapter 2.

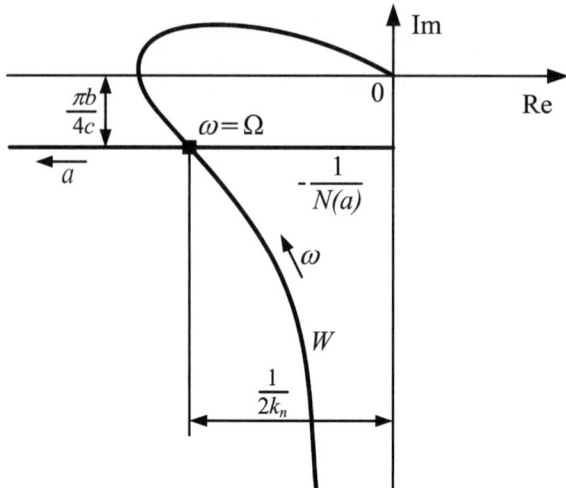

Fig. 1.4. DF analysis of periodic motions

1.5 Asymmetric oscillations in relay servo systems: DF analysis

Now we turn to the analysis of asymmetric oscillations in the relay servo system, which is the key step to the analysis of the system response to constant and slow varying disturbances and reference input signals.

Assume that the input to the system is a constant signal $f_0 : f(t) \equiv f_0$. Then an asymmetric periodic motion occurs in the system (Fig. 1.5), so that each signal now has a periodic and a constant term $u(t) = u_0 + u_p(t), y(t) = y_0 + y_p(t), \sigma(t) = \sigma_0 + \sigma_p(t)$, where the subscript '0' refers to the constant term in the Fourier series, and the subscript 'p' refers to the periodic term of the function (the sum of periodic terms of the Fourier series).

The constant term is the mean or averaged value of the signal over the period. Now let us imagine that we slowly slew the input from a certain negative value to a positive value, so that at each value of the input, the system exhibits a stable oscillation, and measure the values of the constant term of the control (mean control) versus the constant term of the error signal (mean error). By doing this, we can determine the constant term of the control signal as a function of the constant term of the error signal, which is not a discontinuous but a smooth function: $u_0 = u_0(\sigma_0)$.

We call it the *bias function*. Two typical bias functions are depicted in Fig. 1.6. The described effect is known as the *chatter smoothing* phenomenon, which is studied in [56]. The derivative of the mean control with respect to the mean error taken around the point of zero mean error $\sigma_0 = 0$ (corresponding to zero constant input) provides the *equivalent gain* of the relay k_n, which would be similar to the concept of the so-called *incremental gain* [50, 91] of the

1.5 Asymmetric oscillations in relay servo systems: DF analysis   15

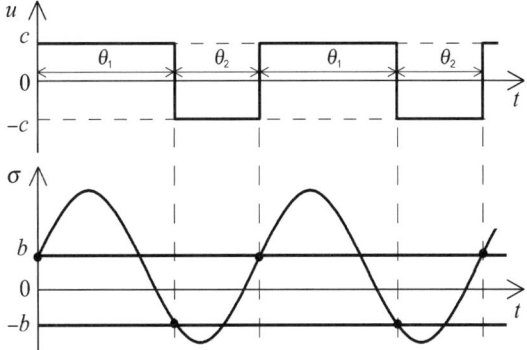

Fig. 1.5. Asymmetric oscillations at unequally spaced switches

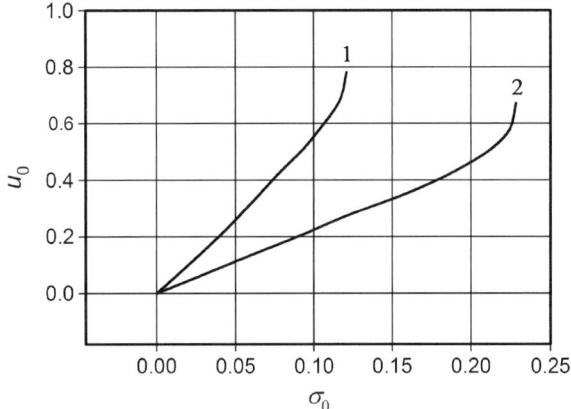

Fig. 1.6. Bias functions ($c = 1$)

describing function method. The *equivalent gain* of the relay is used as a local approximation of the bias function:

$$k_n = du_0/d\sigma_0|_{\sigma_0=0} = \lim_{f_0 \to 0}(u_0/\sigma_0).$$

Because at the slow inputs the relay servo system behaves similar to a linear system with respect to the response to those input signals, finding the *equivalent gain* value is the main point of the input-output analysis. Once it is found, all subsequent analysis of propagation of the slow input signals can be carried out exactly as in a linear system with the relay replaced by the *equivalent gain*. The model obtained via the replacement of the relay with the *equivalent gain* would represent the model of the averaged (over the period of the oscillations) motions in the system. This is especially pertinent to a SM analysis because the deviations of the sliding variable from the zero value are usually small, and the *equivalent gain*, being a local approximation of the *bias function*, usually provides good accuracy. The model obtained as described above is a non-reduced order model. It retains the order of the original system. The non-reduced-order model is described in detail in Chapter 4.

Now let us carry out analysis of asymmetric oscillations in the system Fig. 1.1 caused by a non-zero constant input $f(t) \equiv f_0 \neq 0$. The DF of the hysteretic relay with a biased sine input is represented by the following well-known formula [8]:

$$N(a, \sigma_0) = \frac{2c}{\pi a}\left[\sqrt{1 - \left(\frac{b+\sigma_0}{a}\right)^2} + \sqrt{1 - \left(\frac{b-\sigma_o}{a}\right)^2}\right] - j\frac{4cb}{\pi a^2}, (a \geq b+|\sigma_0|), \tag{1.21}$$

where a is the amplitude of the oscillations. The mean control as a function of a and σ_0 is given by the following formula:

$$u_0(a, \sigma_0) = \frac{c}{\pi}\left(\arcsin\frac{b+\sigma_0}{a} - \arcsin\frac{b-\sigma_0}{a}\right). \tag{1.22}$$

From (1.21) and (1.22), we can obtain the DF of the relay and the derivative of the mean control with respect to the mean error for the symmetric sine input:

$$k_{n(DF)} = \left.\frac{\partial u_0}{\partial \sigma_0}\right|_{\sigma_0=0} = \frac{2c}{\pi a}\frac{1}{\sqrt{1-\left(\frac{b}{a}\right)^2}}. \tag{1.23}$$

1.6 Slow signal propagation through a relay servo system

With the presented methodology of analysis of the effect of constant input, we can now consider analysis of slow signal propagation through a relay servo system. Assume that signals f_0, σ_0, y_0, previously considered constant, are slowly changing signals in comparison to the periodic motions. We will call them the slow components of the motion. By comparatively slow, we mean signals that meet the following condition: those signals can be considered constant over the period of the self-excited oscillations without significant loss of accuracy of the oscillation estimation. Although this is not a rigorous definition, it outlines a framework for the following analysis.

It is also worth noting that due to the feedback action, the system always tries to decrease the value of the error signal σ. This is also true with respect to the averaged value (or slow varying component) of the error signal σ_0. As a result, the averaged value of the error signal normally stays within the linear zone of the bias function (Fig. 1.6). In that case, the periodic solution will not be significantly different from the periodic solution of the symmetric periodic process, which follows from the harmonic balance equation (1.19) in which

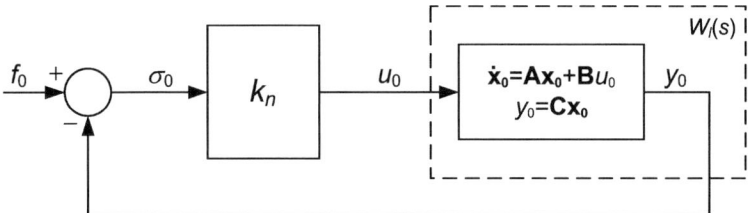

Fig. 1.7. Dynamics of the slow motions

either the DF for the symmetric oscillation (1.18) or the DF for the biased oscillation (1.21) must be used. One notices that if the hysteresis value b of the relay is *zero*, then the frequency of the periodic motions is the same for the symmetric and biased oscillations.[3] The equivalent gain value will be equal to the one found for the infinitesimally small constant input. The only difference between this and the analysis of the response to the constant inputs is the effect of the dynamics of the linear part, which must be accounted for. The dynamics of the slow (averaged over the period of self-excited oscillations) motions can be represented by the block diagram (Fig. 1.7).

The system in Fig. 1.7 is linear, and the dynamics of the slow motions in the relay feedback system are governed by linear equations — due to the chatter smoothing of the relay nonlinearity. It is worth mentioning here that the hysteretic relay does not introduce any lag effect to the slow component of the motion. However, it introduces some dynamics into the fast component of the motion, which follows from the expression for the DF (1.18) having a negative imaginary part.

1.7 Conclusions

In this chapter we outline the frequency-domain approach to analysis of self-excited oscillations and external signal propagation. This approach is based on the describing function method and the harmonic balance concept. Also, the relationship between frequency-domain analysis and the state-space representation is shown via the notion of periodic signal mapping. We show how analysis of external signal propagation can be done via analysis of asymmetric oscillations, and that the oscillations can be analyzed by finding a fixed point of the periodic signal mapping. We introduce notions of the bias function and of the equivalent gain of the relay. These notions are extensively used in subsequent chapters of the book, where an exact method of analysis is presented.

---

[3] This is only true within the framework of the DF analysis.

2

The locus of a perturbed relay system (LPRS) theory

2.1 Introduction to the LPRS

As we considered in the previous chapter, the motions in relay servo systems are normally analyzed as motions in two separate dynamic subsystems: the "slow" subsystem and the "fast" subsystem. The "fast" subsystem pertains to self-excited oscillations or periodic motions. The "slow" subsystem deals with forced motions caused by an input signal or by a disturbance, a non-zero initial conditions component of the motion, and usually pertains to the averaged (over the period of the self-excited oscillation) motion. The two dynamic subsystems interact with each other via a set of parameters: the results of the solution of the "fast" subsystem are used by the "slow" subsystem. This decomposition of the dynamics is possible if the external input is much slower than the self-excited oscillations, which is normally the case. Exactly as in the DF method, we shall proceed from the assumption that the external signals applied to the system are slow in comparison to the oscillations.

Consider again the harmonic balance equation (1.19). Using the formulas for the negative reciprocal of the DF (1.20) and the equivalent gain of the relay (1.23), we can rewrite formula (1.19) as follows:

$$W_l(j\Omega) = -\frac{1}{2}\frac{1}{k_{n(DF)}} + j\frac{\pi}{4c}y_{(DF)}(0). \tag{2.1}$$

In the imaginary part of (2.1), we view the condition of the switch of the relay from minus to plus (defined as zero time) as the equality of the system output to the negative half hysteresis ($-b$): $y_{(DF)}(t=0) = -b$. It follows from (1.21), (1.23), and (2.1) that the frequency of the oscillations and the equivalent gain in the system (1.16) can be varied by changing the hysteresis value $2b$ of the relay. Therefore, the following two mappings can be considered: $M_1 : b \to \Omega$, $M_2 : b \to k_n$. Assume that M_1 has an inverse mapping (it follows from (1.21), (1.23), and (2.1) for the DF analysis and is proved below by deriving an analytical formula) $M_1^{-1} : \Omega \to b$. Applying the chain rule, consider the mapping $M_2\left(M_1^{-1}\right) : \Omega \to b \to k_n$. Now let us define a certain function J as the expression on the right-hand side of formula (2.1) with the additional requirement that the values of the equivalent gain and the output at zero time should be exact values. Applying the mapping $M_2\left(M_1^{-1}\right) : \omega \to b \to k_n$, $\omega \in [0;\infty)$, in which we treat the frequency ω as an independent parameter, we get the following for J:

$$J(\omega) = -\frac{1}{2}\frac{1}{k_n} + j\frac{\pi}{4c}\, y(t)|_{t=0} \qquad (2.2)$$

where $k_n = M_2\left(M_1^{-1}(\omega)\right)$, $y(t)|_{t=0} = M_1^{-1}(\omega)$, $t = 0$ is the time of the switch of the relay from "$-c$" to "$+c$." Thus, $J(\omega)$ comprises the two mappings and is defined as a characteristic of the response of the linear part to the unequally spaced pulse input $u(t)$, subject to $f_0 \to 0$ as the frequency ω varies. The real part of $J(\omega)$ contains information about gain k_n, and the imaginary part of $J(\omega)$ comprises the condition of the switching of the relay and, consequently, contains information about the frequency of the oscillations. By deriving the function that satisfies the above requirements, we can obtain exact values of the frequency of the oscillations and the *equivalent gain*.

We call the function $J(\omega)$ defined above, along with its plot on the complex plane (with the frequency ω varied), the locus of a perturbed relay system (LPRS). Suppose we have computed the LPRS of a given system. Then (as in the DF analysis) we can determine the frequency of the oscillations (as well as the amplitude) and the equivalent gain k_n (Fig. 2.1). The point of intersection of the LPRS and the straight line, which lies at the distance $\pi b/(4c)$ below (if $b > 0$) or above (if $b < 0$) the horizontal axis and parallel to it (line "$-\pi b/4c$"), offers computing the frequency of the oscillations and the equivalent gain k_n of the relay.

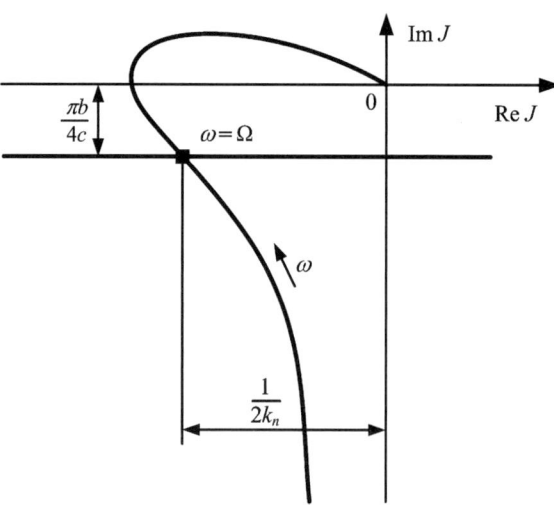

Fig. 2.1. The LPRS and oscillation analysis

According to (2.2), the frequency Ω of the oscillations can be computed by solving the equation:
$$\text{Im } J(\Omega) = -\frac{\pi b}{4c}, \qquad (2.3)$$
(i.e., $y(0) = -b$ is the condition of the relay switch) and the gain k_n can be computed as:
$$k_n = -\frac{1}{2\text{Re } J(\Omega)}. \qquad (2.4)$$

Formula (2.3) provides a periodic solution and is, therefore, a *necessary condition for the existence of a periodic motion* in the system.[1] Formula (2.2) is only a definition and not intended for the purpose of computing the LPRS $J(\omega)$. It is shown below that although $J(\omega)$ is defined through the parameters of the oscillations in a closed-loop system, it can be easily derived from the parameters of the linear part without employing the variables of formula (2.2).

2.2 Computing the LPRS for a non-integrating plant

2.2.1 Matrix state-space description approach

Deriving the computing formula of the LPRS involves only the parameters of the linear part for the case of the non-integrating (self-regulating) linear part given by the matrix differential equations. Let the system be described by the equations (1.16), where $\mathbf{A}$ is nonsingular.

Let us find the periodic solution of system (1.16) at the unequally spaced relay switching caused by a non-zero constant input signal f_0. A common way to find a periodic solution is to use a Poincaré map. Because the control switches are unequally spaced and the oscillations are not symmetric, a Poincaré return map must be considered. Suppose that an asymmetric periodic process of the period T exists in the system. Then, considering the solution for the constant control u,
$$\mathbf{x}(t) = e^{\mathbf{A}t}\mathbf{x}(0) + \mathbf{A}^{-1}(e^{\mathbf{A}t} - \mathbf{I})\mathbf{B}u,$$
the periodic solution of system (1.16) for the control $u = \pm 1$ (it will be shown below that the LPRS is a characteristic of the linear part only and we can assume without loss of generality $c = 1$) can be written as
$$\eta = e^{\mathbf{A}\theta_1}\rho + \mathbf{A}^{-1}(e^{\mathbf{A}\theta_1} - \mathbf{I})\mathbf{B}, \qquad (2.5)$$
$$\rho = e^{\mathbf{A}\theta_2}\eta - \mathbf{A}^{-1}(e^{\mathbf{A}\theta_2} - \mathbf{I})\mathbf{B}, \qquad (2.6)$$

---

[1] The actual existence of a periodic motion depends on a number of other factors, too, including orbital stability of the obtained periodic solution and initial conditions.

where $\rho = \mathbf{x}(0) = \mathbf{x}(T)$, $\eta = \mathbf{x}(\theta_1)$, for the periodic solution, and θ_1, θ_2 are the positive and the negative pulse durations of the periodic control $u(t)$. Formulas (2.5) and (2.6) are Poincaré return maps for the system (sequential numbers of switches are not shown). The periodic solution of system (1.16) can be obtained through finding a fixed point of the Poincaré return map (solution of (2.5) and (2.6)), which is given as follows:

$$\rho = (\mathbf{I} - e^{\mathbf{A}T})^{-1}\mathbf{A}^{-1}[e^{\mathbf{A}T} - 2e^{\mathbf{A}\theta_2} + \mathbf{I}]\mathbf{B}, \tag{2.7}$$

$$\eta = (\mathbf{I} - e^{\mathbf{A}T})^{-1}\mathbf{A}^{-1}[2e^{\mathbf{A}\theta_1} - e^{\mathbf{A}T} - \mathbf{I}]\mathbf{B}. \tag{2.8}$$

We now need to consider the periodic solution (2.7) and (2.8) as a result of the feedback action. The conditions of the switches of the relay can be written as

$$\begin{aligned} f_0 - y(0) &= b \\ f_0 - y(\theta_1) &= -b. \end{aligned} \tag{2.9}$$

Having solved the set of equations (2.9) for f_0, we obtain: $f_0 = (y(0)+y(\theta_1))/2$. Hence, the constant term of $\sigma(t)$ is

$$\sigma_0 = f_0 - y_0 = (y(0) + y(\theta_1))/2 - y_0, \tag{2.10}$$

and the real part of the LPRS definition formula can be transformed into

$$\operatorname{Re} J(\omega) = -0.5 \lim_{\gamma \to \frac{1}{2}} \frac{0.5[y(0) + y(\theta_1)] - y_0}{u_0}, \tag{2.11}$$

where $\gamma = \frac{\theta_1}{\theta_1 + \theta_2} = \frac{\theta_1}{T}$. Then $\theta_1 = \gamma T$, $\theta_2 = (1-\gamma)T$, $u_0 = 2\gamma - 1$, and (2.11) can be written as

$$\operatorname{Re} J(\omega) = -0.5 \lim_{\gamma \to \frac{1}{2}} \frac{0.5\mathbf{C}[\rho + \eta] - y_0}{2\gamma - 1},$$

where ρ and η are given by (2.7) and (2.8), respectively. The imaginary part of the definition formula of $J(\omega)$ can be transformed into:

$$\operatorname{Im} J(\omega) = \frac{\pi}{4}\mathbf{C} \lim_{\gamma \to \frac{1}{2}} \rho.$$

Finally, the state-space description–based formula of the LPRS can be derived on the basis of the previous two formulas and (2.7), (2.8) as follows:

$$\begin{aligned} J(\omega) &= -0.5\mathbf{C}[A^{-1} + \tfrac{2\pi}{\omega}(\mathbf{I} - e^{\frac{2\pi}{\omega}\mathbf{A}})^{-1}e^{\frac{\pi}{\omega}\mathbf{A}}]\mathbf{B} \\ &\quad + j\tfrac{\pi}{4}\mathbf{C}(\mathbf{I} + e^{\frac{\pi}{\omega}\mathbf{A}})^{-1}(\mathbf{I} - e^{\frac{\pi}{\omega}\mathbf{A}})\mathbf{A}^{-1}\mathbf{B}. \end{aligned} \tag{2.12}$$

Therefore, if the system is given in the state-space form (1.16), then formula (2.12) can be used to compute the LPRS. The LPRS computed as (2.12) comprises all possible periodic solutions and equivalent gain values for a given linear part. For that reason, the LPRS is a relatively universal frequency-domain characteristic of the linear part of a relay servo system. An actual periodic solution for a given linear part and parameters of the relay can be found from equation (2.3). A detailed derivation of the LPRS is given in the Appendix.

The subroutine "lprsmatr" (see Appendix) can be used for the LPRS computing per formula (2.12).

2.2.2 Partial fraction expansion technique

We now derive the LPRS formula when the description of the linear part is given in the form of the transfer function expanded into partial fractions. We first prove the additivity property of the LPRS $J(\omega)$.

Theorem 2.1. *(additivity property).* If the transfer function $W_l(s)$ of the linear part is a sum of n transfer functions $W_l(s) = W_1(s) + W_2(s) + ... + W_n(s)$, then the LPRS $J(\omega)$ can be calculated as a sum of n LPRS: $J(\omega) = J_1(\omega) + J_2(\omega) + ... + J_n(\omega)$, where $J_i(\omega)$ $(i = 1, ..., n)$ is the LPRS of the relay system with the transfer function of the linear part being $W_i(s)$.

Proof. We prove the property for $n = 2$: if the property is true for $n = 2$, it is true for any n. Consider the steady asymmetric oscillations in the system when $f(t) \equiv f_0 \neq 0$. Assume that a unimodal asymmetric limit cycle occurs (Fig. 1.4). Suppose that the frequency Ω of the oscillations is known, and the control amplitude c, as well as the pulse duration (θ_1 and θ_2) of the periodic control $u(t)$, are given. If $W_l(s) = W_1(s) + W_2(s)$, then the output is $y(t) = y_1(t) + y_2(t)$, where $y_i(t), i = \overline{1,2}$ is the output of the linear part, which has the transfer function $W_i(s), i = \overline{1,2}$ with its input $u(t)$ as above. Substitute $y_1(t) + y_2(t)$ for $y(t)$ in (2.10) and obtain $\sigma_0 = \sigma_{01} + \sigma_{02}$, where $\sigma_{01} = (y_1(0) + y_1(\theta_1))/2 - y_{01}$, $\sigma_{02} = (y_2(0) + y_2(\theta_1))/2 - y_{02}$, y_{01} and y_{02} are the constant terms of $y_1(t)$ and $y_2(t)$, respectively. Thus, when the parameters of $u(t)$ are as specified above, the constant term of $\sigma(t)$ is equal to the sum of the constant terms of $\sigma_1(t)$ and $\sigma_2(t)$ where $\sigma_1(t)$ and $\sigma_2(t)$ are the errors in two different relay systems with the transfer functions $W_1(s)$ and $W_2(s)$, respectively. Because the additivity property is true for σ_0, it is also true for σ_0/u_0 because u_0 is constant and, consequently, this is true for $\lim(\sigma_0/u_0)$. It is also obvious that $y(0) = y_1(0) + y_2(0)$. Thus, according to (2.2): $J(\omega) = J_1(\omega) + J_2(\omega)$. ∎

The additivity property offers a way of computing the LPRS $J(\omega)$ via expanding $W_l(s)$ into the sum of first- and second-order dynamics (partial fractions), calculating the component LPRS $J_i(\omega)$ for each of them, and summing the LPRS $J_i(\omega)$. Analytical formulas for $J(\omega)$ of first- and second-order dynamics are derived in this chapter below and presented in Table 2.1. The respective MATLAB functions are given in the Appendix (functions "lprs1ord," "lprsint," "lprs2ord1," "lprs2ord2," "lprs2ord3," "lprs2ord4," and "lprsfopdt").

Table 2.1. Formulas of the LPRS $J(\omega)$

Tr. fun. $W(s)$	The LPRS $J(\omega)$
$\frac{K}{s}$	$0 - j\frac{\pi^2 K}{8\omega}$
$\frac{K}{Ts+1}$	$\frac{K}{2}(1 - \alpha \operatorname{csch} \alpha) - j\frac{\pi K}{4}\tanh(\alpha/2)$, $\alpha = \pi/(T\omega)$
$\frac{K}{(T_1 s+1)(T_2 s+1)}$	$\frac{K}{2}[1 - T_1/(T_1 - T_2)\alpha_1 \operatorname{csch} \alpha_1 - T_2/(T_2 - T_1)\alpha_2 \operatorname{csch} \alpha_2)]$ $-j\frac{\pi K}{4}/(T_1 - T_2)[T_1\tanh(\alpha_1/2) - T_2\tanh(\alpha_2/2)]$, $\alpha_1 = \pi/(T_1\omega), \quad \alpha_2 = \pi/(T_2\omega)$
$\frac{K}{s^2+2\xi s+1}$	$\frac{K}{2}[(1 - (B + \gamma C)/(\sin^2\beta + \sinh^2\alpha)]$ $-j\frac{\pi K}{4}(\sinh\alpha - \gamma\sin\beta)/(\cosh\alpha + \cos\beta)$, $\alpha = \pi\xi/\omega, \quad \beta = \pi(1-\xi^2)^{1/2}/\omega, \quad \gamma = \alpha/\beta$ $B = \alpha\cos\beta\sinh\alpha + \beta\sin\beta\cosh\alpha,$ $C = \alpha\sin\beta\cosh\alpha - \beta\cos\beta\sinh\alpha$
$\frac{Ks}{s^2+2\xi s+1}$	$\frac{K}{2}[\xi(B + \gamma C) - \pi/\omega\cos\beta\sinh\alpha]/(\sin^2\beta + \sinh^2\alpha)]$ $-j\frac{\pi K}{4}(1-\xi^2)^{-1/2}\sin\beta/(\cosh\alpha + \cos\beta)$, $\alpha = \pi\xi/\omega, \quad \beta = \pi(1-\xi^2)^{1/2}/\omega, \quad \gamma = \alpha/\beta$ $B = \alpha\cos\beta\sinh\alpha + \beta\sin\beta\cosh\alpha,$ $C = \alpha\sin\beta\cosh\alpha - \beta\cos\beta\sinh\alpha$
$\frac{Ks}{(s+1)^2}$	$\frac{K}{2}[\alpha(-\sinh\alpha + \alpha\cosh\alpha)/\sinh^2\alpha - j0.25\pi\alpha/(1+\cosh\alpha)]$, $\alpha = \pi/\omega$
$\frac{Ks}{(T_1 s+1)(T_2 s+1)}$	$\frac{K}{2}/(T_2 - T_1)[\alpha_2\operatorname{csch}\alpha_2 - \alpha_1\operatorname{csch}\alpha_1]$ $-j\frac{\pi K}{4}/(T_2 - T_1)[\tanh(\alpha_1/2) - \tanh(\alpha_2/2)]$, $\alpha_1 = \pi/(T_1\omega), \quad \alpha_2 = \pi/(T_2\omega)$
$\frac{Ke^{-\tau s}}{Ts+1}$	$\frac{K}{2}(1 - \alpha e^\gamma \operatorname{csch}\alpha) + j\frac{\pi K}{4}\left(\frac{2e^{-\alpha}e^\gamma}{1+e^{-\alpha}} - 1\right)$, $\alpha = \frac{\pi}{T\omega}, \quad \gamma = \frac{\tau}{T}$

2.2.3 Transfer function description approach

Another formula for $J(w)$ can now be derived for the case of the linear part given by a transfer function. Suppose the linear part does not have integrators. We write the Fourier series expansion of the signal $u(t)$ (Fig. 1.5)

$$u(t) = u_0 + 4c/\pi \sum_{k=1}^{\infty} \sin(\pi k\theta_1/(\theta_1 + \theta_2))/k \times \{\cos(k\omega\theta_1/2)\cos(k\omega t) + \sin(k\omega\theta_1/2)\sin(k\omega t)\},$$

where $u_0 = c(\theta_1 - \theta_2)/(\theta_1 + \theta_2)$, $\omega = 2\pi/(\theta_1 + \theta_2)$. Therefore, $y(t)$ as a response of the linear part with the transfer function $W_l(s)$ can be written as

$$y(t) = y_0 + 4c/\pi \sum_{k=1}^{\infty} \sin(\pi k\theta_1/(\theta_1 + \theta_2))/k \\ \times \{\cos(k\omega\theta_1/2)\cos[k\omega t + \varphi_l(k\omega)] \\ + \sin(k\omega\theta_1/2)\sin[k\omega t + \varphi_l(k\omega)]\}A_l(k\omega), \quad (2.13)$$

2.2 Computing the LPRS for a non-integrating plant

where $\varphi_l(k\omega) = \arg W_l(jk\omega)$, $A_l(k\omega) = |W_l(jk\omega)|$, $y_0 = u_0|W_l(j0)|$. The conditions of the switches of the relay have the form of equations (2.9) where $y(0)$ and $y(\theta_1)$ can be obtained from (2.13) if we set $t = 0$ and $t = \theta_1$, respectively:

$$y(0) = y_0 + 4c/\pi \sum_{k=1}^{\infty}[0.5\sin(2\pi k\theta_1/(\theta_1+\theta_2))\mathrm{Re}W_l(jk\omega) \\ + \sin^2(\pi k\theta_1/(\theta_1+\theta_2))\mathrm{Im}W_l(jk\omega)]/k, \quad (2.14)$$

$$y(\theta_1) = y_0 + 4c/\pi \sum_{k=1}^{\infty}[0.5\sin(2\pi k\theta_1/(\theta_1+\theta_2))\mathrm{Re}W_l(jk\omega) \\ - \sin^2(\pi k\theta_1/(\theta_1+\theta_2))\mathrm{Im}W_l(jk\omega)]/k. \quad (2.15)$$

Differentiating (2.9) with respect to f_0 (and taking into account (2.14) and (2.15)), we obtain the formulas containing the derivatives at the point $\theta_1 = \theta_2 = \theta = \pi/\omega$. Solving those equations for $d(\theta_1 - \theta_2)/df_0$ and $d(\theta_1 + \theta_2)/df_0$, we obtain: $d(\theta_1 + \theta_2)/df_0|_{f_0=0} = 0$, which corresponds to the derivative of the frequency of the oscillations, and:

$$\frac{d(\theta_1 - \theta_2)}{df_0}\bigg|_{f_0=0} = 2\theta/[c(|W_l(0)| + 2\sum_{k=1}^{\infty}\cos(\pi k)\mathrm{Re}W_l(\omega k))]. \quad (2.16)$$

Considering the formula of the closed-loop system transfer function, we can write:

$$\frac{d(\theta_1 - \theta_2)}{df_0}\bigg|_{f_0=0} = k_n/(1 + k_n|A_l(0)|)2\theta/c. \quad (2.17)$$

Solving equations (2.16) and (2.17) together for k_n, we obtain the following expression:

$$k_n = 0.5/\sum_{k=1}^{\infty}(-1)^k \mathrm{Re}W_l(k\pi/\theta). \quad (2.18)$$

Taking into account formula (2.18), the identity $\omega = \pi/\theta$, and the definition of the LPRS (2.2), we obtain the final form of expression for $\mathrm{Re}J(\omega)$. Similarly, solving the set of equations (2.9), where $\theta_1 = \theta_2 = \theta$ and $y(0)$ and $y(\theta_1)$ have the form (2.14) and (2.15), respectively, we obtain the final formula of $\mathrm{Im}J(\omega)$. Putting the real and the imaginary parts together, we obtain the final formula of the LPRS $J(\omega)$ for relay systems with non-integrating plants:

$$J(\omega) = \sum_{k=1}^{\infty}(-1)^{k+1}\mathrm{Re}W_l(k\omega) + j\sum_{k=1}^{\infty}\frac{1}{2k-1}\mathrm{Im}W_l[(2k-1)\omega]. \quad (2.19)$$

The subroutine "lprsser200" (see Appendix) can be used for the LPRS computing per formula (2.19), which takes the sum of 200 terms of the series.

2.2.4 Orbital stability of relay systems

The stability of periodic orbits (limit cycles) is usually referred to as orbital stability. The notion of orbital stability is different from the notion of stability of an equilibrium point: for an orbitally stable motion, the difference between the perturbed and unperturbed motions does not necessarily vanish. What is important is that the perturbed motion in an orbitally stable system converges to the orbit of the unperturbed system. More details about this type of stability are provided in [62]. In relay feedback systems, analysis of orbital stability can be reduced to the analysis of certain equivalent discrete-time systems with time instants corresponding to the switches of the relay, which can be obtained from the original system by the Poincaré map of the motion with an initial perturbation. The stability condition based on this approach was proposed in [2]. If we assume that the initial state is $\mathbf{x}(0) = \rho + \delta\rho$, where $\delta\rho$ is the initial perturbation, and find the mapping $\delta\rho \to \delta\eta$, we can make a conclusion about orbital stability of the system by considering the Jacobian matrix of this mapping. A detailed derivation of the Jacobian matrix that relates the perturbations at switching times is given in the Appendix.

Therefore, the stability criterion can be formulated as follows.

Theorem 2.2. *The relay feedback system (1.16) is locally orbitally asymptotically stable if and only if all eigenvalues of the matrix*

$$\Phi_0 = \left[\mathbf{I} - \frac{\mathbf{v}(\frac{T}{2}-)\mathbf{C}}{\mathbf{C}\mathbf{v}(\frac{T}{2}-)}\right] e^{\mathbf{A}\frac{T}{2}}, \tag{2.20}$$

where $T = \frac{2\pi}{\Omega}$ is the period of the oscillations, $\mathbf{v}$ is the velocity matrix,

$$\mathbf{v}\left(\frac{T}{2}-\right) = 2\left(\mathbf{I} - e^{\mathbf{A}T}\right)^{-1}\left(e^{\mathbf{A}\frac{T}{2}} - e^{\mathbf{A}T}\right)\mathbf{B} = 2\left(\mathbf{I} + e^{\mathbf{A}T/2}\right)^{-1}e^{\mathbf{A}T/2}\mathbf{B},$$

have magnitudes less than one.

In addition to the stability analysis, the direction of the relay switch must be verified, too [94]. This condition is formulated as the following inequality,

$$\dot{y}\left(\frac{T}{2}-\right) = \mathbf{C}\mathbf{v}\left(\frac{T}{2}-\right) > 0,$$

where $\mathbf{v}\left(\frac{T}{2}-\right)$ is given by the previous formula.

2.3 Computing the LPRS for an integrating plant

2.3.1 Matrix state-space description approach

For an integrating linear part, the formulas derived above cannot be used without certain modifications. Despite the fact that the solution $\mathbf{x}(t)$ of the

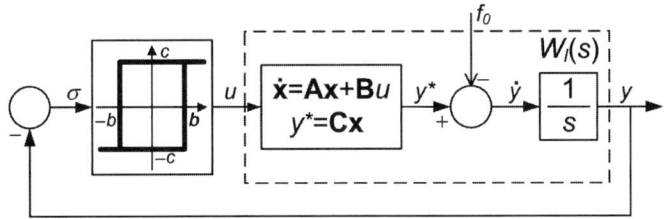

Fig. 2.2. Relay servo system with integrating linear part

system is well-defined even if the matrix **A** does not have an inverse, the above results are not applicable to an integrating linear part. In the case of unequally spaced switches, a system with a conventional description, strictly speaking, cannot have a periodic process even if a ramp signal is applied to the input of the system in Fig. 1.3. The motion occurring in such a system is a combination of a periodic and a ramp motion — due to unlimited integration. To enable the system to have an asymmetric periodic motion, we must transpose the constant input signal to the integrator input (Fig. 2.2). The balance of the constant terms of the signals in the various points of the system must be achieved for periodic motion to occur.

Similarly, we derive the formulas of $J(\omega)$ for the case of an integrating linear part. The state-space description of the system (Fig. 2.2) has the following form,

$$\dot{\mathbf{x}} = \mathbf{A}\mathbf{x} + \mathbf{B}u, \tag{2.21}$$

$$\dot{y} = \mathbf{C}\mathbf{x} - f_0, \tag{2.22}$$

$$u = \begin{cases} +c \text{ if } \sigma = -y \geq b \text{ or } \sigma > -b, u(t-0) = c \\ -c \text{ if } \sigma = -y \leq -b \text{ or } \sigma < b, u(t-0) = -c \end{cases}$$

where $\mathbf{A} \in R^{(n-1)\times(n-1)}, \mathbf{B} \in R^{(n-1)\times 1}, \mathbf{C} \in R^{1\times(n-1)}$, **A** is nonsingular, f_0 is a constant input to the system, σ is the error signal, and $u(t-0)$ is the control value at the time immediately preceding the current time. Note that formula (2.22) defines not the output y but its derivative, which adds an integrator to the linear part. A separate consideration of the variable $y(t)$ from the other state variables is possible due to the integrating property of the linear part. This allows us at first to find a periodic solution for $\mathbf{x}(t)$ (for a given unequally spaced switching), and after that to determine a periodic solution for the system output. The periodic solution for $\mathbf{x}(t)$ is given above (formulas (2.7) and (2.8)). The periodic output $y(t)$ can be obtained by integrating equation (2.22) from the initial states determined by formulas (2.7) and (2.8). As a result, for the control amplitude $c = 1$, the system output can be written as

$$y_1(t) = y_1(0) - \mathbf{C}\mathbf{A}^{-1}\mathbf{B}t - f_0 t + \mathbf{C}\mathbf{A}^{-1}[(e^{\mathbf{A}t} - \mathbf{I})\rho + \mathbf{A}^{-1}(e^{\mathbf{A}t} - \mathbf{I})\mathbf{B}], \tag{2.23}$$

$$y_2(t) = y_1(\theta_1) + \mathbf{C}\mathbf{A}^{-1}\mathbf{B}t - f_0 t + \mathbf{C}\mathbf{A}^{-1}[(e^{\mathbf{A}t} - \mathbf{I})\eta - \mathbf{A}^{-1}(e^{\mathbf{A}t} - \mathbf{I})\mathbf{B}], \tag{2.24}$$

where $y_1(t) = y(t)$, $y_2(t) = y(t+\theta_1)$.

The time t in formulas (2.23) and (2.24) is independent, and $t = 0$ in formula (2.23) is the time of the switch from minus to plus, and in formula (2.24) $t = 0$ is the time of the switch from plus to minus. For periodic motion, the following equations hold, which represents a Poincaré return map:

$$y(\theta_1) = y(0) - (\mathbf{CA}^{-1}\mathbf{B} - f_0)\theta_1 + \mathbf{CA}^{-1}[(e^{\mathbf{A}\theta_1} - \mathbf{I})\rho + \mathbf{A}^{-1}(e^{\mathbf{A}\theta_1} - \mathbf{I})\mathbf{B}], \quad (2.25)$$

$$y(0) = y(\theta_1) + (\mathbf{CA}^{-1}\mathbf{B} - f_0)\theta_2 + \mathbf{CA}^{-1}[(e^{\mathbf{A}\theta_2} - \mathbf{I})\eta - \mathbf{A}^{-1}(e^{\mathbf{A}\theta_2} - \mathbf{I})\mathbf{B}]. \quad (2.26)$$

Analysis of equations (2.25) and (2.26) shows that the set of equations has a solution if and only if

$$f_0 = -\mathbf{CA}^{-1}\mathbf{B}(2\gamma - 1), \quad (2.27)$$

$$\gamma = \frac{\theta_1}{\theta_1 + \theta_2} = \frac{\theta_1}{T},$$

which corresponds to the situation when the constant term of the signal $y^*(t)$ is equal to f_0 and, therefore, the constant term at the integrator input is zero — the only possibility for the system to have a periodic process. Furthermore, equations (2.25) and (2.26) are equivalent and have an infinite number of solutions. To understand why, note that if a periodic signal with zero constant term is applied to the integrator input, its output signal is not uniquely determined, but, depending on the initial value, can represent an infinite number of biased periodic signals. To define a unique solution, introduce an additional condition:

$$y(\theta_1) = -y(0). \quad (2.28)$$

The solution of equations (2.25) and (2.28) results in

$$y(0) = \mathbf{CA}^{-1}\mathbf{B}\gamma(1-\gamma)T + \tfrac{1}{4}\mathbf{CA}^{-2}\{(\mathbf{I} - e^{\mathbf{A}T})^{-1}[6e^{\mathbf{A}T} - 3(e^{\mathbf{A}\theta_1} + e^{\mathbf{A}\theta_2}) - e^{\mathbf{A}T}(e^{\mathbf{A}\theta_1} + e^{\mathbf{A}\theta_2}) + 2\mathbf{I}] - (e^{\mathbf{A}\theta_1} + e^{\mathbf{A}\theta_2}) + 2\mathbf{I}\}\mathbf{B}. \quad (2.29)$$

The output at $t = \theta_1$ is a negative value of the same formula. Thus, we find the periodic solution of system (2.21), (2.22). The LPRS formula can be derived from the analysis of the closed-loop system with an unequally spaced switching control having an infinitesimally small asymmetry. The constant term y_0 of the output $y(t)$ is determined as the sum of integrals of functions (2.23) and (2.24) divided by the period T

$$y_0 = \frac{1}{T}\{\int_0^{\theta_1} y_1(\tau)d\tau + \int_0^{\theta_2} y_2(\tau)d\tau\}, \quad (2.30)$$

where $y_1(\tau)$ is given by (2.23) and $y_2(\tau)$ is given by (2.24). The formula for the real part of $J(\omega)$ can be transformed into

$$\mathrm{Re}J(\omega) = \lim_{\gamma \to \frac{1}{2}} \frac{y_0}{2\gamma - 1}, \tag{2.31}$$

where expression (2.30) can be used for computing y_0. The formula of the imaginary part of $J(\omega)$ is determined by (2.29) with a coefficient, which follows from the LPRS definition. Finally, the LPRS for the case of an integrating linear part can be expressed with the following formula

$$J(\omega) = \tfrac{1}{4}\mathbf{C}\mathbf{A}^{-2}\{(\mathbf{I}-\mathbf{D}^2)^{-1}[\mathbf{D}^2 - (\mathbf{I} + \tfrac{4\pi}{\omega}\mathbf{A})\mathbf{D} + \mathbf{D}^3 - \mathbf{I}] + \mathbf{D} - \mathbf{I}\}\mathbf{B}$$
$$+ j\tfrac{\pi}{8}\mathbf{C}\mathbf{A}^{-1}\{\tfrac{\pi}{\omega} + \mathbf{A}^{-1}[(\mathbf{I}-\mathbf{D}^2)^{-1}(3\mathbf{D}^2 - 3\mathbf{D} - \mathbf{D}^3 + \mathbf{I}) - \mathbf{D} - \mathbf{I}]\}\mathbf{B}, \tag{2.32}$$

where $\mathbf{D} = e^{\frac{\pi}{\omega}\mathbf{A}}$. Therefore, the state-space description–based LPRS formula for the case of an integrating linear part has been derived above.

The subroutine "lprsmatrint" (see Appendix) can be used for the LPRS computing per formula (2.32).

2.3.2 Transfer function description approach

We derive the LPRS formula for the case of an integrating linear part given by a transfer function. The model suitable for the following analysis is given in Fig. 2.2. One can notice that the periodic terms of the signals of the system Fig. 2.2 are the same as the periodic terms of respective signal of the system Fig. 1.3. For that reason, we can use some results of the above analysis for the case of a non-integrating linear part. The constant input f_0 causes an asymmetry in the periodic motion. In a steady periodic motion, the constant term of the input signal to the integrator is zero. Therefore, the constant input is compensated for by the constant term of the signal $y^*(t)$, and the output of the system can again be written as in formula (2.13). However, the value of y_0 in (2.13) is different. Now it does not directly depend on u_0. The values of $y(0)$ and $y(\theta_1)$ are given by formulas (2.14) and (2.15), as before. In other words, the input $\sigma(t)$ to the relay has two terms: the constant term σ_0 and the periodic term $\sigma_p(t)$. The periodic term $\sigma_p(t)$ coincides with that of formula (2.13) (the negative value of the latter), and the constant term is $\sigma_0 = -y_0$. Because the input to the relay does not include the external input f_0, the following equation holds:

$$y(0) + y(\theta_1) = 0.$$

Solving this equation, we find that $\sigma_0 = -y_0$, and

$$\sigma_0 = \frac{2c}{\pi} \sum_{k=1}^{\infty} \sin\left(\frac{2\pi k \theta_1}{\theta_1 + \theta_2}\right) \mathrm{Re}W_l(jk\omega).$$

The equivalent gain k_n can be obtained as a reciprocal of the derivative $d\sigma_0/du_0$ at $\theta_1 = \theta_2 = \pi/\omega$. We compute the following limit

$$\lim_{\gamma \to \frac{1}{2}} \frac{\sigma_0}{u_0} = 2\sum_{k=1}^{\infty}(-1)^k \cdot \mathrm{Re}W_l(jk\omega).$$

The real part of the LPRS is given by $\mathrm{Re}J(\omega) = -0.5/k_n$, where the equivalent gain k_n is the reciprocal of the above limit. The imaginary part of the LPRS remains the same for the case of an integrating linear part. And finally, a formula for the LPRS can be constructed on the basis of the definition (2.2) and the above analysis. The final formula of the LPRS is given as follows:

$$J(\omega) = \sum_{k=1}^{\infty}(-1)^{k+1}\mathrm{Re}W_l(k\omega) + j\sum_{k=1}^{\infty}\frac{1}{2k-1}\mathrm{Im}W_l[(2k-1)\omega]. \quad (2.33)$$

One can see that formula (2.33) coincides with formula (2.19). Therefore, despite the different model and different mechanism of generation of the constant term in the error signal, the LPRS formula expressed in terms of the frequency response of the linear part remains the same. For an accurate calculation of a point of $J(\omega)$, the few first terms of the series (2.33) are enough as a rule. It can be shown that the series (2.33) always converges for strictly proper transfer functions. Formula (2.33) can also be used for the LPRS calculation from a frequency response characteristic (Bode plot, Nyquist plot) of the linear part.

2.3.3 Orbital stability of relay systems

An integrating plant provides significantly different dynamics in comparison with a non-integrating plant. Therefore, the stability conditions in [2] cannot be directly used for stability analysis of the systems with integrating plants. The formal reason is that the matrix $\mathbf{A}$ is not invertible. However, with the plant description as in (2.21), (2.22), the matrix $\mathbf{A}$ refers only to the non-integrating part of the plant and, thus, has an inverse. Again, if we assume that the initial state is $\mathbf{x}(0) = \rho + \delta\rho$, where $\delta\rho$ is the initial perturbation, and find the mapping $\delta\rho \to \delta\eta$, we can make a conclusion about the orbital stability of the system by considering the Jacobian matrix of this mapping. A detailed derivation of the Jacobian matrix that relates the perturbations at switching times is given in the Appendix.

Therefore, the stability criterion can be formulated as follows.

Theorem 2.3. *The relay feedback system (2.21), (2.22) is locally orbitally asymptotically stable if and only if all the eigenvalues of the matrix*

$$\Phi_0 = -\frac{\mathbf{v}\left(\frac{T}{2}-\right)\mathbf{CA}^{-1}(e^{\mathbf{A}\frac{T}{2}} - \mathbf{I})}{\dot{y}_p\left(\frac{T}{2}\right)} + e^{\mathbf{A}\frac{T}{2}}, \quad (2.34)$$

where $T = \frac{2\pi}{\Omega}$ is the period of the oscillations, $\mathbf{v}$ is the velocity matrix,

$$\mathbf{v}(\frac{T}{2}-) = 2\left(\mathbf{I} + e^{\mathbf{A}T/2}\right)^{-1} e^{\mathbf{A}T/2}\mathbf{B},$$

and

$$\dot{y}_p\left(\frac{T}{2}\right) = \mathbf{CA}^{-1}\mathbf{B} - 2\mathbf{CA}^{-1}\left(\mathbf{I} + e^{\mathbf{A}T/2}\right)^{-1}\mathbf{B}$$

have magnitudes less than one.

In addition to the stability analysis, the direction of the relay switch must be verified, too. This condition is formulated as the following inequality:

$$\dot{y}_p\left(\frac{T}{2}\right) > 0,$$

where $\dot{y}_p\left(\frac{T}{2}\right)$ is given by the previous formula.

2.4 Computing the LPRS for a plant with a time delay

2.4.1 Matrix state-space description approach

Consider now the linear part with a time delay. Let the plant be

$$\dot{\mathbf{x}} = \mathbf{A}\mathbf{x} + \mathbf{B}u$$
$$y = \mathbf{C}\mathbf{x}$$

and the control

$$u = \begin{cases} +1 \text{ if } \sigma(t-\tau) = f_0 - y(t-\tau) \geq b & \text{or} \quad \sigma(t-\tau) > -b,\ u(t-) = 1 \\ -1 \text{ if } \sigma(t-\tau) = f_0 - y(t-\tau) \leq -b & \text{or} \quad \sigma(t-\tau) < b,\ u(t-) = -1 \end{cases}$$

where $\mathbf{A} \in R^{n \times n}$, $\mathbf{B} \in R^{n \times 1}$, $\mathbf{C} \in R^{1 \times n}$ are matrices, and $\mathbf{A}$ is nonsingular. We note that $t = 0$ corresponds to the time that the error signal reaches the hysteresis values $\sigma = b$, $\dot{\sigma} > 0$. The control $u(t)$ switches from -1 to $+1$ not at time $t = 0$ but at time $t = \tau$. The solution for the constant control $u = \pm 1$ is

$$\mathbf{x}(t) = e^{\mathbf{A}(t-\tau)}\mathbf{x}(\tau) \pm \mathbf{A}^{-1}(e^{\mathbf{A}(t-\tau)} - \mathbf{I})\mathbf{B}, \quad t > \tau.$$

Therefore, also

$$\mathbf{x}(\tau) = e^{\mathbf{A}\tau}\mathbf{x}(0) - \mathbf{A}^{-1}(e^{\mathbf{A}\tau} - \mathbf{I})\mathbf{B}.$$

Denoting $\rho_p = \mathbf{x}(\tau) = \mathbf{x}(T + \tau)$, $\eta_p = \mathbf{x}(\theta_1 + \tau)$, where θ_1 is the length of the positive pulse of control, we can partly use the results obtained above for the linear part without time delay. A detailed derivation of the LPRS for the case being considered is given in the Appendix. The final state-space description–based formula of the LPRS can be written as follows:

$$J(\omega) = -0.5\mathbf{C}\left[\mathbf{A}^{-1} + \frac{2\pi}{\omega}\left(\mathbf{I} - e^{\frac{2\pi}{\omega}\mathbf{A}}\right)^{-1} e^{\left(\frac{\pi}{\omega} - \tau\right)\mathbf{A}}\right]\mathbf{B}$$
$$+ j\frac{\pi}{4}\mathbf{C}\left(\mathbf{I} + e^{\frac{\pi}{\omega}\mathbf{A}}\right)^{-1}\left(\mathbf{I} + e^{\frac{\pi}{\omega}\mathbf{A}} - 2e^{\left(\frac{\pi}{\omega} - \tau\right)\mathbf{A}}\right)\mathbf{A}^{-1}\mathbf{B}. \tag{2.35}$$

The subroutine "lprsmatrdel" (see Appendix) can be used for the LPRS computing per formula (2.35).

2.4.2 Orbital asymptotic stability

Let us extend the above methodology to the case of a plant with a time delay. Consider only the case of symmetric oscillations and apply a simplified approach. Again we need to find the mapping of the initial perturbation (at time $t = 0$) into the perturbation at the time corresponding to the condition $\sigma = -b, \dot{\sigma} < 0$. Denoting the initial perturbation $\delta \mathbf{x}(0)$, we can write the mapping $\delta \mathbf{x}(0) \to \delta \rho$ as

$$\delta \rho = e^{\mathbf{A}\tau}\delta \mathbf{x}(0),$$

and the mapping $\delta \mathbf{x}(0) \to \delta \rho \to \delta \mathbf{x}(T/2)$ as

$$\delta \mathbf{x}(T/2) = e^{\mathbf{A}(T/2-\tau)}\delta \rho = e^{\mathbf{A}(T/2-\tau)}e^{\mathbf{A}\tau}\delta \mathbf{x}(0) = e^{\mathbf{A}T/2}\delta \mathbf{x}(0).$$

We note that the stability condition (2.20) is the product of two multipliers. The first one is a mapping due to the change of the switching instant when the initial perturbation is present, and the second one is the mapping $\delta \mathbf{x}(0) \to \delta \mathbf{x}(T/2)$. The second multiplier comes from above, and the first multiplier stays the same, subject to the formula for the velocity matrix below. Therefore, the stability criterion can be formulated as follows.

Theorem 2.4. *The relay feedback system with a time-delay plant is locally orbitally asymptotically stable if and only if all the eigenvalues of the matrix*

$$\Phi_0 = \left[\mathbf{I} - \frac{\mathbf{v}(\frac{T}{2}-)\mathbf{C}}{\mathbf{C}\mathbf{v}(\frac{T}{2}-)}\right]e^{\mathbf{A}\frac{T}{2}},$$

where $T = \frac{2\pi}{\Omega}$ is the period of the oscillations, $\mathbf{v}$ is the velocity matrix,

$$\mathbf{v}\left(\tfrac{T}{2}-\right) = \dot{\mathbf{x}}\left(\tfrac{T}{2}-\right) = \mathbf{A}\mathbf{x}\left(\tfrac{T}{2}\right) + \mathbf{B} = -\mathbf{A}\mathbf{x}(0) + \mathbf{B}$$
$$= -\mathbf{A}(\mathbf{I} + e^{\mathbf{A}T/2})^{-1}\mathbf{A}^{-1}\left[\mathbf{I} + e^{\mathbf{A}T/2} - 2e^{\mathbf{A}(T/2-\tau)}\right]\mathbf{B} + \mathbf{B}$$
$$= -(\mathbf{I} + e^{\mathbf{A}T/2})^{-1}\left[\mathbf{I} + e^{\mathbf{A}T/2} - 2e^{\mathbf{A}(T/2-\tau)}\right]\mathbf{B} + \mathbf{B}$$
$$= 2\left(\mathbf{I} + e^{\mathbf{A}T/2}\right)^{-1}e^{\mathbf{A}(T/2-\tau)}\mathbf{B}$$

have magnitudes less than one.

In addition to the stability analysis, the direction of the relay switch must be verified, too. This condition is formulated as the following inequality

$$\dot{y}\left(\frac{T}{2}-\right) = \mathbf{Cv}\left(\frac{T}{2}-\right) > 0,$$

where $\mathbf{v}\left(\frac{T}{2}-\right)$ is given by the previous formula.

2.5 LPRS of first-order dynamics

As mentioned above, one of the possible techniques of LPRS computing is to represent the transfer function as partial fractions, compute the LPRS of the component transfer functions (partial fractions), and add those partial LPRS together in accordance with Theorem 2.1. To apply this technique, we have to know the formulas of the LPRS for first- and second-order dynamics. These are of similar meaning and importance as the characteristics of first- and second-order dynamics in linear system analysis.

The knowledge of the LPRS of the low-order dynamics is important for other reasons, too. Some features of the LPRS of low-order dynamics can be extended to higher-order systems. Those features are considered in Chapter 4.

Let us find the formula of the LPRS for the first-order dynamics given by the transfer function $W(s) = K/(Ts+1)$.

We derive an analytical formula for $J(\omega)$, $\omega \in [0; \infty)$. There exist nonsymmetrical oscillations in the system (Fig. 2.3) if $f(t) \equiv f_0 \neq 0$. The system model can be written as the following set of equations:

$$\begin{cases} y(\theta_1) = y(0)\exp(-\theta_1/T) + cK(1 - \exp(-\theta_1/T)) \\ y(\theta_1 + \theta_2) = y(\theta_1)\exp(-\theta_2/T) - cK(1 - \exp(-\theta_2/T)) \\ y(\theta_1 + \theta_2) = y(0) \\ f_0 - y(0) = b \\ f_0 - y(\theta_1) = -b. \end{cases} \quad (2.36)$$

Solving (2.36), we obtain θ_1 and θ_2:

$$\theta_1 = -T\ln(2b/(f_0 - b - cK) + 1), \quad (2.37)$$

$$\theta_2 = -T\ln(1 - 2b/(f_0 + b + cK)). \quad (2.38)$$

Consider the limit $\lim_{f_0 \to 0}(u_0/f_0)$: on the one hand, it can be derived from (2.37), (2.38) (taking into account that $u_0 = c(\theta_1 - \theta_2)/(\theta_1 + \theta_2)$)

$$\lim_{f_0 \to 0}(u_0/f_0) = -2bcT/(\theta(b - cK)(b + cK)) \quad (2.39)$$

where $\theta = \lim_{f_0 \to 0} \theta_{1,2} = -T\ln(1 - 2b/(cK + b))$; on the other hand, it can be related to the gain k_n by the formula of a closed-loop system

$$\lim_{f_0 \to 0} (u_0/f_0) = k_n/(1 + k_n K)). \tag{2.40}$$

A formula for k_n can easily be found from (2.39) and (2.40)

$$k_n = -T(\exp(-2\theta/T) - 1)/[K(2\theta \exp(-\theta/T) + T(\exp(-2\theta/T) - 1))]$$

from which a formula for $\mathrm{Re}J(\omega)$ (where $\omega = \pi/\theta$) can be obtained. An expression for $\mathrm{Im}J(\omega)$ can be found by solving the set of equations (2.36) with $f_0 = 0$. Finally we obtain

$$J(\omega) = \frac{K}{2}\left(1 - \frac{\pi}{T\omega}\operatorname{csch}\frac{\pi}{T\omega}\right) - j\frac{\pi K}{4}\tanh\frac{\pi}{2\omega T}, \tag{2.41}$$

where $\operatorname{csch} x$ and $\tanh x$ are hyperbolic cosecant and tangent, respectively. The subroutine "lprs1ord" (see Appendix) can be used for the LPRS computing per formula (2.41).

The plot of the LPRS for $K = 1$, $T = 1$ is given in Fig. 2.3. The whole plot is totally located in the 4th quadrant. The point $(0.5K; -j\frac{\pi}{4}K)$ corresponds to the frequency $\omega = 0$, and the point $(0; j0)$ corresponds to the frequency $\omega = \infty$. The high-frequency segment of the LPRS has the imaginary axis as an asymptote.

With the formula for the LPRS available, we can easily find the frequency of periodic motions in the relay servo system with the linear part being the first-order dynamics. The LPRS is a continuous function of the frequency, and for every hysteresis value from the range $b \in [0; cK]$, there exists a periodic solution of the frequency that can be determined from (2.3), (2.41), which is

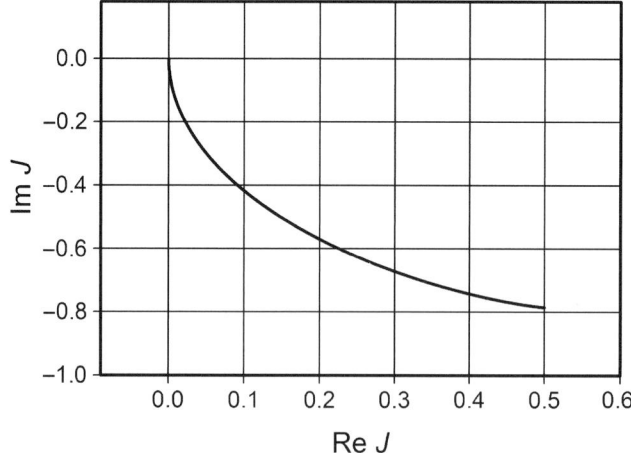

Fig. 2.3. The LPRS of first-order dynamics

$$\Omega = \frac{\pi}{2T} \tanh^{-1}\left(\frac{b}{cK}\right). \tag{2.42}$$

It is easy to show that when the hysteresis value b tends to zero, then the frequency of the periodic solution tends to infinity

$$\lim_{b \to 0} \Omega = \infty,$$

and when the hysteresis value b tends to cK, then the frequency of the periodic solution tends to zero

$$\lim_{b \to cK} \Omega = 0.$$

From (2.41), we can also see that the imaginary part of the LPRS is a monotone function of the frequency. Therefore, the condition of the existence of a finite frequency periodic solution holds for any non-zero hysteresis value from the specified range, and the limit for $b \to 0$ exists and corresponds to infinite frequency.

It is easy to show that the oscillations are always orbitally stable. The stability of a periodic solution is usually verified by finding eigenvalues of the Jacobian of the corresponding Poincaré map [62]. For the first-order system, the only eigenvalue of this Jacobian will always be zero, as there is only one system variable, which also determines the condition of the switch of the relay.

2.6 LPRS of second-order dynamics

Now we carry out a similar analysis for second-order dynamics. Let the matrix $\mathbf{A}$ of (1.16) be $\mathbf{A} = [0\ 1;\ -a_1\ -a_2]$. Here, consider a few cases, all with $a_1 > 0, a_2 > 0$.

A. Let $a_2^2 - 4a_1 < 0$. Then the plant transfer function can be written as:

$$W(s) = K/(T^2 s^2 + 2\xi T s + 1). \tag{2.43}$$

The LPRS formula can be found, for example, by expanding the above transfer function into partial fractions and applying formula (2.41) obtained for the first-order dynamics. However, the coefficients of those partial fractions will be complex numbers, and this circumstance must be considered. The formula of the LPRS for the second-order dynamics given by transfer function (2.43) can be written as follows:

$$J(\omega) = \frac{K}{2}\left(1 - \frac{g + \gamma h}{\sin^2 \beta + \sinh^2 \alpha}\right) - j\frac{\pi K}{4}\frac{\sinh \alpha - \gamma \sin \beta}{\cosh \alpha + \cos \beta} \tag{2.44}$$

where
$$\alpha = \frac{\pi\xi}{\omega T},$$
$$\beta = \frac{\pi\sqrt{1-\xi^2}}{\omega T},$$
$$\gamma = \alpha/\beta$$
$$g = \alpha\cos\beta\sinh\alpha + \beta\sin\beta\cosh\alpha,$$
$$h = \alpha\sin\beta\cosh\alpha + \beta\cos\beta\sinh\alpha.$$

The subroutine "lprs2ord1" (see Appendix) can be used for the LPRS computing per formula (2.44).

The plots of the LPRS for $K = 1$, $T = 1$ and different values of damping factor ξ are given in Fig. 2.4 (#1–$\xi = 1$, #2–$\xi = 0.85$, #3–$\xi = 0.7$, #4–$\xi = 0.55$, #5 – $\xi = 0.4$). The high-frequency segment of the LPRS of the second-order plant approaches the real axis.

Now, with the LPRS formula available, we analyze possible existence of the periodic solution in the relay feedback system with the plant being the second-order dynamics. Consider two limits of $J(\omega)$ that can be obtained from (2.44):

$$\lim_{\omega\to\infty} J(\omega) = (0; j0); \quad \lim_{\omega\to 0} J(\omega) = (0.5K; -j\frac{\pi}{4}K).$$

They give the two boundary points of the LPRS corresponding to zero frequency and infinite frequency. Analysis of function (2.44) shows that it does not have intersections with the real axis except at the origin. Because $J(\omega)$ is a continuous function of the frequency ω (this follows from formula (2.44)), a solution of equation (2.3) exists for any $b \in (0; cK)$. This means that a periodic solution of finite frequency exists for the second-order system for every value

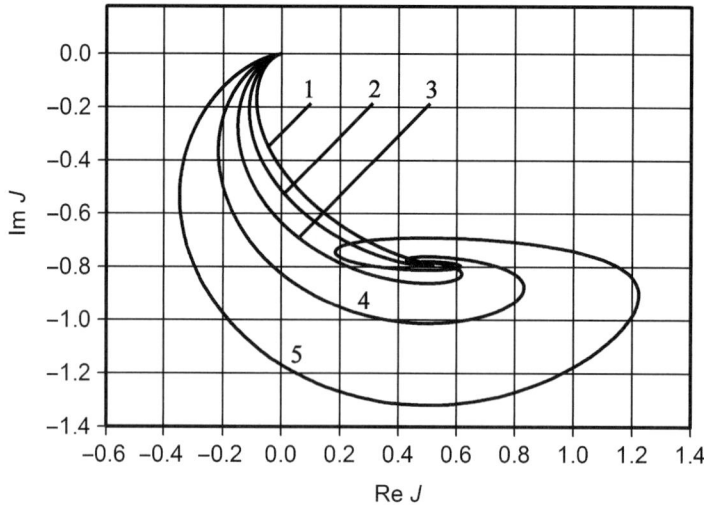

Fig. 2.4. The LPRS of second-order dynamics

of b within the specified range, and there is a periodic solution of infinite frequency for $b = 0$.

Now we analyze the stability of those periodic solutions. We write the Jacobian of the Poincaré map of the relay system:

$$\Phi = \left[\mathbf{I} - \frac{v\mathbf{C}}{\mathbf{C}v}\right] e^{\mathbf{A}\pi/\omega}, \quad (2.45)$$

where $v = 2(\mathbf{I} + e^{\mathbf{A}\pi/\omega})^{-1} e^{\mathbf{A}\pi/\omega} \mathbf{B}$. If all the eigenvalues of the matrix Φ have magnitudes smaller than *one*, the periodic motion is orbitally asymptotically stable. For the second-order system, we obtain analytical formulas of the matrix Φ eigenvalues

$$\lambda_1 = 0,$$
$$\lambda_2 = -a_1 \alpha_1^2 \pi^2 / \omega^2 + \alpha_0(a_2 \alpha_1 \pi / \omega - \alpha_0), \quad (2.46)$$

where

$$\alpha_0 = \frac{\lambda_{1A} \exp(\lambda_{2A}\pi/\omega) - \lambda_{2A} \exp(\lambda_{1A}\pi/\omega)}{\lambda_{1A} - \lambda_{2A}}$$

$$\alpha_1 = \frac{\exp(\lambda_{1A}\pi/\omega) - \exp(\lambda_{2A}\pi/\omega)}{\lambda_{1A} - \lambda_{2A}} \frac{\omega}{\pi}.$$

λ_{1A} and λ_{2A} are eigenvalues of the matrix A,

$$\lambda_{1A} = 0.5(-a_2 + \sqrt{a_2^2 - 4a_1}), \lambda_{2A} = 0.5(-a_2 - \sqrt{a_2^2 - 4a_1}).$$

Therefore, if $|\lambda_2| < 1$, then the periodic solution is stable. From (2.46), we can also find the limit corresponding to the oscillations of infinite frequency, which is of much interest in sliding mode control theory: $\lim_{\omega \to \infty} \lambda_2 = 0$. Therefore, the periodic solution of infinite frequency is stable.

B. Consider the case when $a_2^2 - 4a_1 = 0$. To obtain the LPRS formula, we use formula (2.44) and find the limit for $\xi \to 1$. The LPRS for this case is given in Fig. 2.4 (#1). All subsequent analysis and conclusions are the same as in case A.

C. Assume that $a_2^2 - 4a_1 > 0$. Then the transfer function can be expanded into two partial fractions, and according to Theorem 2.1, the LPRS can be computed as a sum of the two components. The subsequent analysis is similar to the previous one.

D. Assume that $a_1 = 0$. Then the transfer function is $W(s) = K/[s(Ts + 1)]$. For this plant, the LPRS is given by the following formula, which can be obtained via partial fraction expansion of the transfer function expression and application of the LPRS formulas of the first-order dynamics:

$$J(\omega) = \frac{K}{2}\left(\frac{\pi}{T\omega} \operatorname{csch} \frac{\pi}{T\omega} - 1\right) + j\frac{\pi K}{4}\left(\tanh \frac{\pi}{2\omega T} + \frac{\pi}{2\omega}\right). \quad (2.47)$$

The plot of the LPRS for $K = 1$, $T = 1$ is given in Fig. 2.5. The whole plot is totally located in the 3rd quadrant. The point $(0.5K; -j\infty)$ corresponds

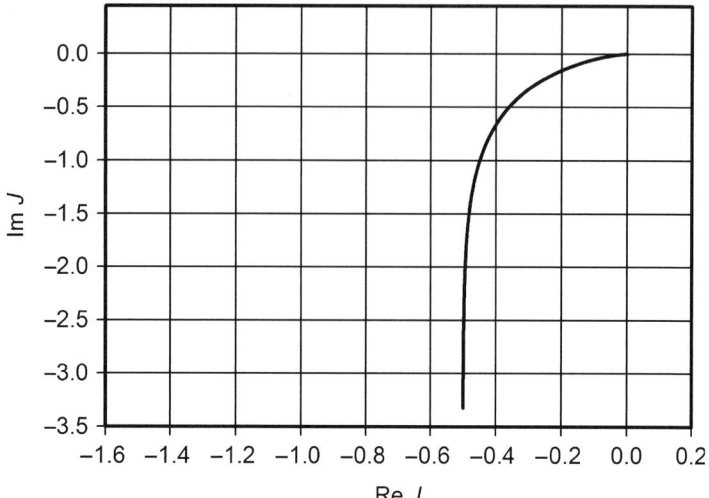

Fig. 2.5. The LPRS of integrating second-order plant

to the frequency $\omega = 0$, and the point $(0; j0)$ corresponds to the frequency $\omega = \infty$. The high-frequency segment of the LPRS has the real axis as an asymptote.

Again, applying the LPRS formula and the same approach, we can prove that the periodic solution of the relay feedback system, with the plant being the second-order dynamics, exists; that in the case of the ideal relay it is the oscillations of infinite frequency; and that the periodic solution is orbitally asymptotically stable.

2.7 LPRS of first-order plus dead-time dynamics

Many industrial processes can be adequately approximated by the first-order plus time-delay transfer function

$$W(s) = \frac{Ke^{-\tau s}}{Ts + 1} \tag{2.48}$$

where K is the process gain, T is a time constant, and τ is a time delay (dead time). This factor results in the particular importance of the analysis of these dynamics. To apply the above idea to the process (2.48), we need to obtain the formula of the LPRS for the transfer function (2.48).

Consider the equation of the periodic process with unequally spaced switching in the relay feedback system (Fig. 1.5) with the plant being the transfer function (2.48). At first, for an auxiliary purpose, we find a response of the first-order plant without time delay to the steady periodic pulse control of the amplitude c, with positive pulse length θ_1 and negative

2.7 LPRS of first-order plus dead-time dynamics

pulse length θ_2. The steady periodic response of such a plant can be described by the following expressions:

$$y^*(\theta_1) = y^*(0) \cdot e^{-\theta_1/T} + cK(1 - e^{-\theta_1/T}) \qquad (2.49)$$

$$y^*(0) = y^*(\theta_1) \cdot e^{-\theta_2/T} + cK(1 - e^{-\theta_2/T}). \qquad (2.50)$$

Formulas (2.49) and (2.50) are a Poincaré return map for the feedback relay system with the plant being a first-order transfer function. Solution of (2.49) and (2.50) provides the following result:

$$y_{\min} = y^*(0) = cK \frac{2e^{-\theta_2/T} - e^{-(\theta_1+\theta_2)/T} - 1}{1 - e^{-(\theta_1+\theta_2)/T}} \qquad (2.51)$$

$$y_{\max} = y^*(\theta_1) = cK \frac{1 + e^{-(\theta_1+\theta_2)/T} - 2e^{-\theta_2/T}}{1 - e^{-(\theta_1+\theta_2)/T}}. \qquad (2.52)$$

Denote the values of the output at the switching instants $y_{\min}$ and $y_{\max}$ (2.51) and (2.52). With $y_{\min}$ and $y_{\max}$ available, we can now write the equations of the asymmetric periodic process in the system with the first-order plus dead-time plant:

$$y(\theta_1) = y_{\min} \cdot e^{-(\theta_1-\tau)/T} + cK(1 - e^{-(\theta_1-\tau)/T}) \qquad (2.53)$$

$$y(0) = y_{\max} \cdot e^{-(\theta_2-\tau)/T} - cK(1 - e^{-(\theta_2-\tau)/T}) \qquad (2.54)$$

$$f_0 - y(0) = b \qquad (2.55)$$

$$f_0 - y(\theta_1) = -b. \qquad (2.56)$$

First, we derive the formula of the imaginary part of the LPRS for the given plant. According to the definition, the imaginary part of the LPRS is the value of the system output at the time of the switch from "−" to "+." Because the input f_0 tends to zero, to derive the formula of the imaginary part we consider the symmetric oscillations. In that case $y(\theta_1) = -y(0)$, and the solution of equations (2.53)–(2.56) is fairly straightforward:

$$\lim_{f_0 \to 0} y(0) = cK \left(\frac{2e^{-\alpha} \cdot e^{\gamma}}{1 + e^{-\alpha}} - 1 \right) \qquad (2.57)$$

where

$$\alpha = \frac{\theta}{T} = \frac{\pi}{T\omega} \text{ and } \gamma = \frac{\tau}{T}.$$

Now we derive the formula of the real part of the LPRS for the given plant. We solve equations (2.53)–(2.56) for θ_1 and θ_2:

$$\theta_1 = -T \ln \frac{f_0 + b - cK}{f_0 - b + cK - 2cKe^{\gamma}} \qquad (2.58)$$

$$\theta_2 = -T \ln \frac{f_0 - b + cK}{2cKe^\gamma + f_0 + b - cK}. \tag{2.59}$$

We find the limiting value of the positive and negative pulse length for $f_0 \to 0$:

$$\lim_{f_0 \to 0} \theta_1 = \lim_{f_0 \to 0} \theta_2 = \theta = T \cdot \ln \frac{cK(2e^\gamma - 1) + b}{cK - b}. \tag{2.60}$$

In formula (2.60), θ is half of the period of the symmetric oscillations. Consequently, the frequency of the oscillations is: $\Omega = \pi/\theta$.

Now we derive a formula of $\lim_{f_0 \to 0} \frac{\theta_1 - \theta_2}{f_0}$. It can be derived from (2.58) and (2.59) but it must not contain b or f_0 on the right-hand side. For that reason, formula (2.60) is helpful. After a number of transformations, we obtain:

$$\lim_{f_0 \to 0} \frac{\theta_1 - \theta_2}{f_0} = \frac{T(1 + e^{-\alpha}) \cdot (1 - e^{-\alpha})}{cKe^\gamma \cdot e^{-\alpha}}. \tag{2.61}$$

Formula (2.61) does not contain b or f_0 in the right-hand side.

Taking into account the relation between θ_1, θ_2 and u_0, we obtain the following limit:

$$\lim_{f_0 \to 0} \frac{u_0}{f_0} = \frac{c}{2\theta} \cdot \lim_{f_0 \to 0} \frac{\theta_1 - \theta_2}{f_0} = \frac{(1 + e^{-\alpha}) \cdot (1 - e^{-\alpha})}{2\alpha Ke^\gamma \cdot e^{-\alpha}}. \tag{2.62}$$

Another expression for the same limit is the formula of the closed-loop system that uses the equivalent gain of the relay k_n:

$$\lim_{f_0 \to 0} \frac{u_0}{f_0} = \frac{k_n}{1 + k_n K}. \tag{2.63}$$

Equating the right-hand sides of (2.62) and (2.63), we obtain the equation for the equivalent gain k_n. After solving it, we obtain the formula of the real part of the LPRS (taking into account the fact that the real part is the reciprocal of the equivalent gain with the coefficient -0.5). Finally, we put together the real and the imaginary parts and obtain the formula for the LPRS for the first-order plus dead-time transfer function as follows:

$$J(\omega) = \frac{K}{2}(1 - \alpha e^\gamma \operatorname{csch} \alpha) + j\frac{\pi}{4} K \left(\frac{2e^{-\alpha} e^\gamma}{1 + e^{-\alpha}} - 1 \right). \tag{2.64}$$

The subroutine "lprsfopdt" (see Appendix) can be used for the LPRS computing per formula (2.64).

Let us compute the LPRS and plot it for various values of γ. The plots of the LPRS for $\gamma = 0$ (#1), $\gamma = 0.2$ (#2), $\gamma = 0.5$ (#3), $\gamma = 1.0$ (#4), and $\gamma = 1.5$ (#5) are depicted in Fig. 2.6. All the plots begin at the point $(0.5, -j\pi/4)$ that corresponds to the frequency $\omega = 0$. Plot number 1 (which corresponds to zero dead-time) comes to the origin, which corresponds to

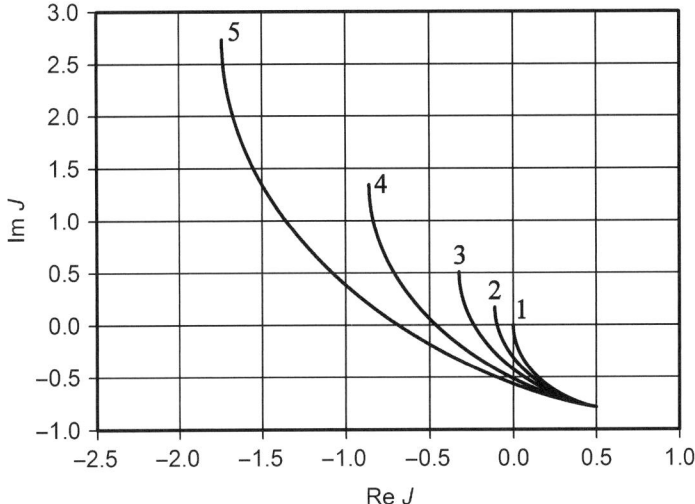

Fig. 2.6. The LPRS of first-order plus dead-time dynamics

infinite frequency. Other plots are defined only for frequencies less than those corresponding to half of the period. Therefore, they do not reach the origin.

Formula (2.64) can be validated by computing the LPRS for the same values of γ as in Fig. 2.6, with the use of the series expression (2.19). An application of formula (2.19) to the transfer function (2.48) provides the same results as formula (2.64).

2.8 Some properties of the LPRS

The knowledge of certain properties of the LPRS is computationally helpful, especially for the design of linear compensators with the use of the LPRS method. One of these properties, probably the most important, was formulated in Theorem 2.1: it is the additivity property. A few other properties relating to the boundary points corresponding to zero frequency and infinite frequency are considered below.

Consider a non-integrating linear part of the relay servo system given by equations (1.16). We find the coordinates of the initial point of the LPRS corresponding to zero frequency. For that purpose, let us find the limit of function $J(\omega)$ for ω tending to zero. Using formula (2.12), we can write:

$$\lim_{\omega \to 0} J(\omega) =$$
$$\mathbf{C} \lim_{\omega \to 0} \left\{ -0.5[\mathbf{A}^{-1} + \tfrac{2\pi}{\omega}(\mathbf{I} - e^{\frac{2\pi}{\omega}\mathbf{A}})^{-1} e^{\frac{\pi}{\omega}\mathbf{A}}] + j\tfrac{\pi}{4}(\mathbf{I} + e^{\frac{\pi}{\omega}\mathbf{A}})^{-1}(\mathbf{I} - e^{\frac{\pi}{\omega}\mathbf{A}})\mathbf{A}^{-1} \right\} \mathbf{B}.$$

We evaluate the following two limits:

$$\lim_{\omega\to 0}\left[\tfrac{2\pi}{\omega}((\mathbf{I}-e^{\frac{2\pi}{\omega}\mathbf{A}})^{-1}e^{\frac{\pi}{\omega}\mathbf{A}}\right] = \lim_{\omega\to 0}\left[\tfrac{2\pi}{\omega}e^{-\frac{2\pi}{\omega}\mathbf{A}}e^{\frac{\pi}{\omega}\mathbf{A}}\right] = \lim_{\omega\to 0}\left[\tfrac{2\pi}{\omega}e^{-\frac{\pi}{\omega}\mathbf{A}}\right] = \mathbf{0}$$

$$\lim_{\omega\to 0}\left[(\mathbf{I}+e^{\frac{\pi}{\omega}\mathbf{A}})^{-1}(\mathbf{I}-e^{\frac{\pi}{\omega}\mathbf{A}})\right] = \lim_{\omega\to 0}\left[e^{-\frac{\pi}{\omega}\mathbf{A}}e^{\frac{\pi}{\omega}\mathbf{A}}\right] = \mathbf{I}.$$

With these two limits, we can write the limit for the LPRS as follows:

$$\lim_{\omega\to 0} J(\omega) = \left[-0.5 + j\frac{\pi}{4}\right]\mathbf{CA}^{-1}\mathbf{B}. \tag{2.65}$$

The product of matrices $\mathbf{CA}^{-1}\mathbf{B}$ in (2.65) is the negative value of the gain of the plant transfer function. We have thus proved that for a *non-integrating linear part of the relay servo system, the initial point of the corresponding LPRS is $(0.5K; -j\pi/4K)$, where K is the static gain of the linear part.* This coincides with the above analysis of the LPRS of the first- and second-order dynamics: see, for example, Fig. 2.3 and Fig. 2.4.

To find the limit of $J(\omega)$ as ω tends to infinity, consider the following power series expansion of the exponential function.

$$\lim_{\omega\to\infty}\exp\left(\frac{\pi}{\omega}\mathbf{A}\right) = \lim_{\omega\to\infty}\sum_{n=0}^{\infty}\frac{(\pi/\omega)^n}{n!}\mathbf{A}^n = \mathbf{I} + \lim_{\omega\to\infty}\sum_{n=1}^{\infty}\frac{(\pi/\omega)^n}{n!}\mathbf{A}^n = \mathbf{I},$$

and another limit:

$$\lim_{\omega\to\infty}\left\{\tfrac{2\pi}{\omega}\left[\mathbf{I}-\exp\left(\tfrac{2\pi}{\omega}\mathbf{A}\right)\right]^{-1}\right\} = \lim_{\lambda\to 0}\{\lambda[\mathbf{I}-\exp(\lambda\mathbf{A})]^{-1}\}$$

$$= \lim_{\lambda\to 0}\left\{\left(\tfrac{\partial\lambda}{\partial\lambda}\right)\left[\tfrac{\partial(\mathbf{I}-\exp(\lambda\mathbf{A}))}{\partial\lambda}\right]^{-1}\right\} = -\mathbf{A}^{-1}.$$

Finally, taking account of the above two limits, we can prove that the final point of the LPRS for the non-integrating linear part is the origin:

$$\lim_{\omega\to\infty} J(\omega) = 0 + j0. \tag{2.66}$$

Reasoning along the same lines, we can obtain the initial and final points of the LPRS for integrating linear parts, which are as follows:

$$\lim_{\omega\to 0} J(\omega) = 0.5\mathbf{CA}^{-1}\mathbf{B} - j\infty \tag{2.67}$$

$$\lim_{\omega\to\infty} J(\omega) = 0 + j0. \tag{2.68}$$

Some further investigation of asymptotic behavior of the LPRS is done in Chapter 4, which is devoted to analysis of sliding mode control systems. It is shown there that the location of the high-frequency segment of the LPRS determines whether chattering or ideal sliding mode occurs in the system. Here we only consider a few rules that may be helpful for LPRS computing and plotting, as well as verifying calculations of the LPRS.

2.9 LPRS of nonlinear plants

2.9.1 Additivity property

In all the previous sections, we considered relay servo systems with linear plants only. This is a limitation of the aforementioned method. However, the LPRS is a characteristic of the relay servo system that remains meaningful and useful even if the plant is nonlinear. Of course, the same methods of computing cannot be used for nonlinear plants. However, other techniques of computing can be developed on the basis of some properties considered below. The application of this approach is demonstrated in the chapter devoted to analysis and design of the pneumatic servomechanism.

Consider the relay feedback system depicted in Fig. 2.7. Let the system be given by the following equations,

$$\dot{\mathbf{x}} = \mathbf{g}(\mathbf{x}, u),$$
$$y = h(\mathbf{x}), \tag{2.69}$$

$$u = \begin{cases} +c \text{ if } \sigma = f_0 - y \geq b \text{ or } \sigma > -b, u(t-0) = c \\ -c \text{ if } \sigma = f_0 - y \leq -b \text{ or } \sigma < b, u(t-0) = -c, \end{cases}$$

where $\mathbf{g}$ and h are nonlinear functions. We limit our analysis to static symmetric nonlinearities. Assume as before that in the autonomous mode, a symmetric periodic motion exists in the system, and if a non-zero external input is applied to the system, then a periodic motion with unequally spaced switches of the relay occurs (see Fig. 1.5). We use the same definition of the LPRS that was introduced above — except now we have the system with a nonlinear plant

$$J(\omega) = -\frac{1}{2} \lim_{f_0 \to 0} \frac{\sigma_0}{u_0} + j\frac{\pi}{4c} \lim_{f_0 \to 0} y(t)|_{t=0}, \tag{2.70}$$

where $t = 0$ is the time of the switch of the relay from "$-c$" to "$+c$." The definition in formula (2.70) does not require that the plant necessarily be linear. Therefore, if via some technique, we compute and plot the LPRS for this nonlinear plant that satisfies the given definition, then we can calculate the frequency of possible periodic motions and the equivalent gain of the relay

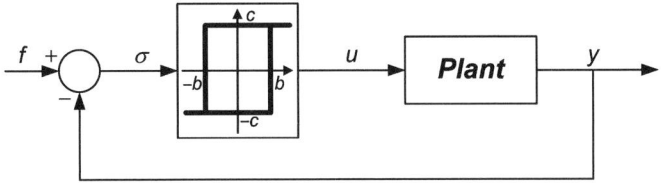

Fig. 2.7. Relay servo system with a nonlinear plant

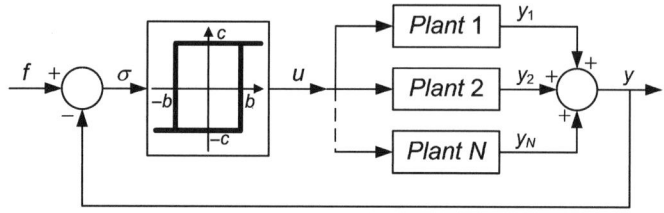

Fig. 2.8. Multichannel plant system

for input-output analysis. Nonlinear plants in the general case do not provide any means for the LPRS computing other than a direct application of formula (2.70). Yet, some features of the LPRS and the system under consideration allow for a simpler approach. The problem is, therefore, to develop a technique or techniques that use those features and do not directly involve the variables of formula (2.70), but rather use the parameters of the plant. Consider the main feature that allows us to simplify the task of LPRS computing.

Assume that the plant can be represented as a number of parallel channels (plant components) as depicted in Fig. 2.8 with each component satisfying the above requirements for the plant. Then the following property is valid.

Theorem 2.5. *(Additivity property of the LPRS). If the plant of the relay servo system can be represented as a sum of N nonlinear (in a general case) plants as depicted in Fig. 2.8,*

$$\dot{\mathbf{x}}_i = \mathbf{g}_i(\mathbf{x}_i, u),$$
$$y_i = h_i(\mathbf{x}_i), \quad i = \overline{1, N},$$
$$y = \sum_{i=1}^{N} y_i,$$

each satisfying the assumptions of equations (2.69), then the LPRS of this system is equal to the sum of N LPRS, each of which corresponds to the relay servo system (Fig. 2.7) with the plant being the plant component of the original system given by i-th equation (Plant 1, Plant 2, Plant N in Fig. 2.8)

$$J(\omega) = J_1(\omega) + J_2(\omega) + \ldots + J_N(\omega), \tag{2.71}$$

where $J_i(\omega)$ is the LPRS for i-th plant.

Proof. Suppose the system in Fig. 2.8 is a type 0 servo system (has a non-integrating plant), a constant input $f(t) \equiv f_0$ is applied, and a limit cycle of frequency Ω does occur. Then the output $y(t)$ at the switching time is $y(0) = y^+$ (switching from "−" to "+") and $y(\theta_1) = y^-$ (switching from "−" to "+").

Each variable $y_i(t)$ ($i = \overline{1, N}$) at the switching time is equal to $y_i(0) = y_i^+$ and $y_i(\theta_1) = y_i^-$. The signals in each system have the following constant terms (mean values): $\sigma_0, u_0, y_0, y_{0i}(i = \overline{1, N})$. The following identities obviously hold:

$$\begin{cases} \sum_{i=1}^{N} y_i^+ = y^+, \\ \sum_{i=1}^{N} y_i^- = y^-, \\ \sum_{i=1}^{N} y_{0i} = y_0. \end{cases} \quad (2.72)$$

The periodic solution of the system in Fig. 2.8 (the switching conditions) can be described as follows:

$$\begin{cases} f_0 - y^+ = b, \\ f_0 - y^- = -b \end{cases} \quad (2.73)$$

Because each signal $y_i(t)$ ($i = \overline{1, N}$) is a periodic function, the periodic solution for each plant component in the relay system in Fig. 2.7 (the plant is supposed to be i-th component of the original plant) exists if the input f_0 and hysteresis value b are equal to

$$f_{0i} = \frac{1}{2}(y_i^+ + y_i^-) \quad (2.74)$$

$$b_{0i} = \frac{1}{2}(y_i^- + y_i^+) \quad (2.75)$$

respectively, which is a solution of system (2.73) for i-th component. Note that the constant inputs and the hysteresis values are different for each of the systems with an i-th plant. Therefore, a periodic solution for the system in Fig. 2.7 with the plant being a plant component from the system in Fig. 2.8 exists, and the output of this system coincides with the component output $y_i(t)$ in the system in Fig. 2.8. The relay controls are identical,

$$u_i(t) = u_2(t) = \ldots = u_N(t) = u(t),$$

which is also true with respect to the constant terms (mean values):

$$u_{01} = u_{02} = \ldots = u_{0N} = u_0. \quad (2.76)$$

Then the following equality holds:

$$\begin{aligned}
\sum_{i=1}^{N} \frac{\sigma_{0i}}{u_{0i}} &= \frac{1}{u_0} \sum_{i=1}^{N} \sigma_{0i} = \frac{1}{u_0} \sum_{i=1}^{N} (f_{0i} - y_{0i}) \\
&= \frac{1}{u_0} \sum_{i=1}^{N} \left[\frac{1}{2}(y_i^+ + y_i^-) - y_{0i} \right] \\
&= \frac{1}{u_0} \left[\frac{1}{2}(y^+ + y^-) - y_0 \right] \\
&= \frac{f_0 - y_0}{u_0} = \frac{\sigma_0}{u_0}.
\end{aligned} \quad (2.77)$$

Formula (2.77) holds for any given frequency $\omega = \Omega$. This means that (2.71) holds with respect to the real part of the LPRS. It is also valid with respect to the imaginary part of the LPRS that directly follows from (2.72). ∎

It should be noted that (2.72) and (2.77) provide an additivity property of the LPRS that is valid not only for infinitesimally small constant terms but also for any finite values. This is very important as we are going to use this property for numerical computing of the LPRS for nonlinear plants.

The proved property suggests some techniques for LPRS computing. The above definition implies existence of the LPRS only within the frequency range where periodic solution is possible. Sometimes it is necessary to calculate the LPRS beyond the frequency range where a periodic solution exists. A typical example of this occurs when the desired frequency of the oscillations is beyond the range of possible frequencies of the oscillations in a non-compensated system.

The main property of the LPRS provides a solution to this problem. If the LPRS definition does not allow for LPRS calculation at the frequency of interest Ω, then additional dynamics with known LPRS can be connected in parallel with the given plant to allow oscillations of frequency Ω to exist in the system. The LPRS of the original plant can be calculated as the LPRS of the latter system minus the LPRS of the known dynamics (at frequency Ω).

Therefore, if the LPRS cannot be calculated at a certain frequency of interest Ω with the use of the LPRS definition, and if by adding components in parallel with the plant (or deleting parallel components) the frequency of interest Ω can be generated, then the LPRS of the given plant at frequency Ω can be calculated as a difference (sum) of the LPRS of the consolidated plant and the LPRS of the components connected in parallel with the given plant; this follows from the additivity property. In the case of linear dynamics connected in parallel with the plant, the LPRS of the linear dynamics can be calculated through the formulas presented in Table 2.1. Thus, the calculation of the LPRS of nonlinear plants depends on how well the auxiliary parallel components are chosen to obtain a necessary frequency of the oscillations.

A typical application of this property is the calculation of the LPRS of a nonlinear plant with two integrators. A periodic solution in the relay system with such a plant doesn't exist (despite the formula of $J(\omega)$ for corresponding linear plant). On the other hand, if second-order dynamics are connected in parallel with such a plant, a periodic solution may exist and the LPRS can be calculated (the switching frequency can be varied by changing the hysteresis value and the parameters of the parallel component).

2.9.2 The LPRS extended definition and open-loop LPRS computing

The LPRS is a plant response to the asymmetric square-wave pulse input signal. It can also be easily seen from formulas (2.12), (2.19), and (2.32) that the LPRS is a function of the plant parameters only. This means that it is

2.9 LPRS of nonlinear plants

possible to construct a certain definition (different from the original definition of the LPRS) that would involve an open-loop consideration of the plant.

Suppose the plant is a type 0 servo system (non-integrating) and the control is a square-wave pulse signal of frequency w, relative pulse duration γ, and amplitude c. Then, the coefficients of the Fourier series of the steady periodic output of the plant can be computed. Relative pulse duration γ is a quotient of positive pulse duration θ_1 and the period of oscillations. Under those assumptions, the plant output is

$$y(t) = y_0 + \sum_{k=1}^{\infty} \{a_k \cdot \cos(kwt) + b_k \cdot \sin(kwt)\}, \qquad (2.78)$$

where y_0, a_k, b_k are the coefficients of the Fourier series, and $t = 0$ is the time of the control switching from "$-c$" to "$+c$."

It follows from (2.78) that at the switching times, the following equalities hold:

$$y^+ = y_0 + \sum_{k=1}^{\infty} a_k, \qquad (2.79)$$

$$y^- = y_0 + \sum_{k=1}^{\infty} \{a_k \cdot \cos(2\pi k\gamma) + b_k \cdot \sin(2\pi k\gamma)\}. \qquad (2.80)$$

Consider the following hypothetical experiment. Suppose that the plant is closed by the feedback (see the system in Fig. 2.7) but the error signal link is disconnected from the relay input and a periodic asymmetric signal from an external generator is introduced to the relay input instead. This results in a square-wave pulse signal of frequency w, relative pulse duration γ, and amplitude c from the relay. At a certain time, we instantaneously disconnect the relay input from the external generator and connect it to the adder output. To allow the oscillations to remain in the closed-loop system, the system input f_0 and relay hysteresis b must satisfy equations (2.73), from which f_0 and b values can be obtained (formulas (2.74) and (2.75)). This shows the existence of a certain equivalence between the open-loop and closed-loop generation of a periodic motion.

Simultaneous consideration of (2.70), (2.74), and (2.75) results in the formula for the LPRS based on the plant output spectrum in the open-loop experiment

$$J(\omega) = -\lim_{\gamma \to \frac{1}{2}} \frac{0.5[y^+ + y^-] - y_0}{2u_0} + j\frac{\pi}{4c} \lim_{\gamma \to \frac{1}{2}} \frac{y^+ - y^-}{2} \qquad (2.81)$$

where $u_0 = c(2\gamma - 1)$, y^+, and y^- are defined by (2.79) and (2.80), respectively.

In practice, the values of γ close to 0.5 can be used for LPRS calculation as per (2.81). If the system is of non-zero type (integrating plant), the adjustments stated above should be made.

We call formula (2.81) the *extended or open-loop definition of the LPRS* because it allows for the LPRS to be defined at frequencies that may not be the frequencies of the oscillations in the closed-loop system (Fig. 2.7). As per (2.81), the LPRS of a given plant can be computed at any frequency.

We call the set of coefficients of the Fourier expansion of plant output y_0, a_k, b_k ($k = 1, ..., m$) at a given frequency ω the *spectral characteristic of the plant at frequency ω*. We have shown above (see the sections devoted to obtaining the LPRS formula from the plant transfer function) that the spectral characteristic at any given frequency can be transformed into the LPRS at the same frequency. In some cases, it is very convenient to calculate and memorize the spectral characteristic. If, for example, a linear compensator is connected in series with the plant (in the error signal), the resulting spectral characteristic is easily calculated as a propagation of the plant output spectrum through the linear compensator. This technique is convenient for the design of a linear compensator. In the chapter devoted to pneumatic servomechanism analysis and design, this technique is further investigated.

2.10 Application of periodic signal mapping to computing the LPRS of some special nonlinear plants

Periodic signal mapping was introduced in Chapter 1, and the relation between this mapping and Poincaré mapping was reviewed above. Now, once the open-loop definition of the LPRS is available, we can apply the periodic signal mapping technique to the open-loop definition and design the methodology for computing the LPRS for nonlinear plants. However, this type of analysis differs from the LPRS analysis in the case of a linear plant. Moreover, according to formula (2.81), the oscillation (the control signal) must be asymmetric to enable us to compute the LPRS. Therefore, we have to consider asymmetric signals and the Fourier series with non-zero constant term, which is a more complex problem compared to the one solved above. Yet, for the purpose of computing the LPRS, the frequency of the control signal can be considered a known value. Also, we can assume a certain small (but sufficient for computing the real part of the LPRS) asymmetry of the control signal. Therefore, we can write the following spectral representation of the control:

$$\mathbf{Q}_u = [q_{u0} \; q_{u1} \; q_{u2} \; q_{u3} \ldots] \tag{2.82}$$

where $q_{u0} = u_0 = c(\theta_1 - \theta_2)/(\theta_1 + \theta_2)$, $\omega = 2\pi/(\theta_1 + \theta_2)$,

$$q_{uk} = \tfrac{4c}{\pi} \sin(\pi k\theta_1/(\theta_1 + \theta_2))/k \times \{\cos(k\omega\theta_1/2)\cos(k\omega t)$$
$$+ \sin(k\omega\theta_1/2)\sin(k\omega t)\}, k = \overline{1, \infty}.$$

In comparison to formula (1.8), formula (2.82) has a constant term. The mapping via linear dynamics given by the transferfunction $W(s)$ is given by the following element-by-element multiplication formula,

$$\mathbf{Q}_y = \mathbf{Q}_u \bullet \mathbf{S} \tag{2.83}$$

where y is the output of these linear dynamics (which can be an internal variable of the plant), and

$$\mathbf{S} = [W(j0)\ W(j\omega)\ W(j2\omega)\ W(j3\omega)\ldots] \tag{2.84}$$

assuming that $W(s)$ is non-integrating.

We consider a few nonlinear models of the plant: the Hammerstein model, the Wiener model, and a nonlinearity preceded and followed by linear dynamics.

The Hammerstein model is given by a cascade connection of a single-valued memoryless nonlinearity followed by linear dynamics (Fig. 2.9). We limit our discussion to single-valued symmetric nonlinearities.

Because the control $u(t)$ is either "$+c$" or "$-c$," the input nonlinearity of the Hammerstein model changes only the value of the effective control amplitude. Instead of "$\pm c$," we have to use "$\pm c_1$," where $c_1 = g(c)$, where the input nonlinearity of the Hammerstein model is $g(u)$. Therefore, the LPRS is not affected by the presence of the input single-valued symmetric nonlinearity of the plant. Yet, when determining the frequency of the periodic motion and the equivalent gain value, the modified value of the relay amplitude must be used instead of the original one.

In the same way, we analyze the effect of the output nonlinearity of the Wiener model. The Wiener model is given by a cascade connection of linear dynamics followed by a single-valued memoryless nonlinearity (Fig. 2.10). Again let us consider only symmetric nonlinearities.

We derive the formula of the LPRS for the plant given by the Wiener model (Fig. 2.10). We rewrite (2.70) in the following form using the notation for the figure,

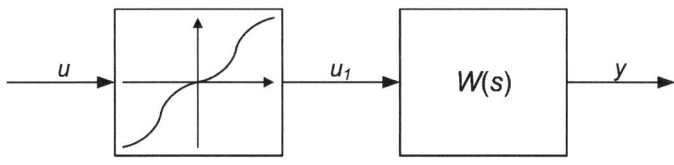

Fig. 2.9. Hammerstein system

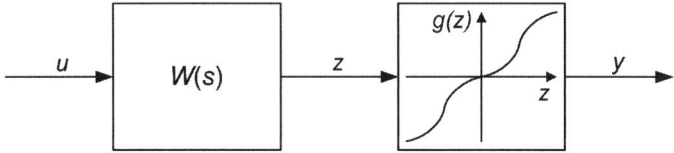

Fig. 2.10. Wiener system

50   2 The LPRS theory

$$\begin{aligned}J(\omega) &= -\frac{1}{2}\lim_{f_0\to 0}\frac{f_0-y_0}{u_0}+j\frac{\pi}{4c}\lim_{f_0\to 0}y(t)\Big|_{t=0}\\ &=-\frac{1}{2}\left(\frac{df_0}{du_0}-\frac{dy_0}{du_0}\right)+j\frac{\pi}{4c}y(t)|_{t=0} \qquad (2.85)\\ &=-\frac{1}{2}\left(\frac{df_0}{du_0}-\frac{dg}{dz}\Big|_{z=0}\frac{dz_0}{du_0}\right)+j\frac{\pi}{4c}g\left(z(t)|_{t=0}\right),\end{aligned}$$

where $...|_{z=0}$ denotes the derivative at the point $z=0$. It follows from formula (2.85) that the LPRS of the Wiener system plant (Fig. 2.10) can be computed as follows,

$$J(\omega) = \frac{dg}{dz}\Big|_{z=0}\mathrm{Re}J_l(\omega)+j\mathrm{Im}J_l(\omega) \qquad (2.86)$$

where $J_l(\omega)$ is the LPRS computed for the linear part only (as per (2.12), (2.32), (2.19)), and the hysteresis value is modified as follows,

$$b_1 = g^{-1}(b), \qquad (2.87)$$

where $g^{-1}(y)$ is the inverse function with respect to the function $g(z)$.

The Hammerstein and Wiener models provide examples of simple nonlinear plants to which the LPRS method can be applied with minimal modifications. However, in real applications, plants dynamics often feature nonlinearities preceded by and followed by linear dynamics as shown in Fig. 2.11. We note that other types of connections between the plant nonlinearity and linear dynamics can be brought to the configuration in Fig. 2.11. We can write the following model of control signal propagation through the nonlinear plant. Assume that the control is given by (2.82). Then the following holds,

$$\mathbf{Q}_{y1} = \mathbf{Q}_u \bullet \mathbf{S_1}, \qquad (2.88)$$

where y_1 is the output of the linear dynamics with the transfer function $W_1(s)$, and $\mathbf{Q}_{y1}$ is the spectrum matrix of $y_1(t)$,

$$\mathbf{S_1} = [W_1(j0)\ W_1(j\omega)\ W_1(j2\omega)\ W_1(j3\omega)\ldots]. \qquad (2.89)$$

The time-domain form of y_1 is obtained as the inverse Fourier transform of its spectral representation:

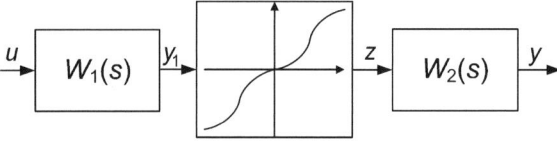

Fig. 2.11. Nonlinearity preceded and followed by linear dynamics

2.10 LPRS of some special nonlinear plants

$$y_1(t) = q_{y10} + 2\sum_{k=1}^{\infty} |q_{y1k}| \cos(\omega t + \arg q_{y1k}). \quad (2.90)$$

As a result, the output of the nonlinearity in the time domain is

$$z(t) = g(y_1(t)). \quad (2.91)$$

In the spectral domain it is as follows:

$$\mathbf{Q}_z = [q_{z0} \; q_{z1} \; q_{z3} \; q_{z5} \ldots] \quad (2.92)$$

where

$$q_{zk} = \frac{1}{T} \int_{-T/2}^{T/2} z(t) [\cos(2k-1)\omega t - j\sin(2k-1)\omega t] \, dt, k = \overline{1, \infty}, \quad (2.93)$$

$$q_{z0} = \frac{1}{T} \int_{-T/2}^{T/2} z(t) dt. \quad (2.94)$$

Subsequently, the output of the plant is given by the following spectral representation,

$$\mathbf{Q}_y = \mathbf{Q}_z \bullet \mathbf{S}_2, \quad (2.95)$$

where

$$\mathbf{S}_2 = [W_2(j0) \; W_2(j\omega) \; W_2(j3\omega) \; W_2(j5\omega) \ldots] \quad (2.96)$$

and in the time domain

$$y(t) = q_{y0} + 2\sum_{k=1}^{\infty} |q_{yk}| \cos(\omega t + \arg q_{yk}), \quad (2.97)$$

where q_{yk} are Fourier coefficients that comprise matrix $\mathbf{Q}_y$. Therefore, utilizing the open-loop computing formula of the LPRS (2.81), the following approximate finite difference schema can be designed for computing the LPRS of the nonlinear plant Fig. 2.11.

$$J(\omega) \approx -\frac{0.5[y^+ + y^-] - y_0}{2u_0} + j\frac{\pi}{4c}\frac{y^+ - y^-}{2}, \quad (2.98)$$

where $y^+ = y(0)$, $y^- = y(\theta_1)$, which is given by (2.97), $y_0 = q_{y0}$, and u_0 is selected to be small enough for the finite difference estimate of the LPRS (2.98) to be precise. Values of u_0 that are too high result in errors in the evaluation of the equivalent gain, and values that are too small may not provide a sufficient resolution of the finite difference approach. As a rule of thumb, u_0 should correspond to the range of expected relative pulse duration values of the control pulses in the system under an external input.

2.11 Comparison of the LPRS with other methods of analysis of relay systems

The LPRS method deals with a classic problem that is examined in many sources. Naturally, the LPRS method should overlap with other existing methods and produce the same results under certain circumstances. Thus it is interesting to compare the LPRS and other methods. The closest two methods are the describing function method and Tsypkin's method.

The describing function method ([8, 50]): Because the DF method is based upon the filtering hypothesis, one might expect that the LPRS method provides the same result under this hypothesis. Indeed, if only the first terms of the series (2.19) of the real and imaginary parts are used (in accordance with the filtering hypothesis), this formula coincides with that of the DF method. The LPRS method, therefore, provides a more precise model of the oscillations and of the input-output properties of a relay system compared to the DF method. In particular, it takes into account the non-sinusoidal shape of the output signal, and the precision enhancement is due to that. If the actual shape of the output signal is close to sinusoidal, both methods provide similar results. Another difference is that the LPRS method does not require harmonic balance conditions to be fulfilled in the closed-loop system; it can handle systems where this condition is not fulfilled (i.e., a system consisting of a hysteretic relay and a first-order linear part or sliding mode control systems).

Tsypkin's method ([94]). The main similarity between Tsypkin's method and the LPRS is in the imaginary parts of the two loci. The imaginary part of the Tsypkin locus is defined as the output value in a periodic motion at the time of the relay switch from minus to plus. The imaginary part of the LPRS is essentially the same: the difference is only in the coefficient. However, the real part of the Tsypkin locus is defined as the derivative of the output at the time of the switch ($t = 0-$) and is intended for verifying the condition of the proper direction of the switch. The real part of the LPRS is defined as a ratio of the two infinitesimally small constant terms of the signals caused by the infinitesimally small asymmetry of switching in a closed-loop system. As a result, the Tsypkin locus is a method of analysis of possible periodic motions only; whereas the LPRS is intended for more complex analysis: the solution of the periodic problem and input-output analysis (disturbance rejection and external signal propagation).

A brief example demonstrates some aspects of the above comparison. Let the plant be the first-order plus dead-time transfer function: $W_l(s) = 0.5\exp(-0.5s)/(1.5s+1)$. The frequency of the periodic solution found via application of the LPRS and of the Tsypkin locus is $\Omega_{LPRS} = \Omega_{Ts} = 3.593\text{s}^{-1}$ (exact value); the same frequency found via application of the DF method is $\Omega_{DF} = 3.516\text{s}^{-1}$ (the error between the two values is 2.1%). The *equivalent gain* values found via application of the LPRS and the DF methods are

$k_{nLPRS} = 6.258$ and $k_{nDF} = 5.371$, respectively. The error between these two values is 14.1%. The true value of the equivalent gain is the same as the one found via the LPRS application. From those results, we can see that even if the DF method may seem precise in terms of the frequency of a periodic solution, the error of the input-output properties may be much larger.

2.12 An example of analysis of oscillations and transfer properties

Let us find a periodic solution and analyze the transfer properties of the relay feedback system with an integrating linear part, given by equations (2.21), (2.22) with the following parameters:

$$\mathbf{A} = \begin{bmatrix} 0 & 1 \\ -0.4 & -0.5 \end{bmatrix}, \quad \mathbf{B} = \begin{bmatrix} 0 \\ 1 \end{bmatrix}, \quad \mathbf{C} = \begin{bmatrix} 1 & 0 \end{bmatrix},$$

$c = 1$, $b = 0.1$. Let an external harmonic signal of frequency 0.01 Hz ($\omega_{in} = 0.0628\text{s}^{-1}$) and amplitude $a_{in} = 20$ be applied to the closed-loop system. We find the frequency and the amplitude of the self-excited oscillation and analyze the system response to the specified external signal.

We compute the LPRS as per formula (2.32) and plot it (Fig. 2.12). For computing the LPRS, the subroutine "lprsmatrint" (see Appendix) can be used. The solution of equation (2.3) corresponding to the point of intersection of the LPRS and the line parallel to the real axis drawn at a distance of $\pi b/(4c) = 0.0785$ below provides the frequency of the oscillations: $\Omega = 0.625\text{s}^{-1}$. The amplitude of the self-excited oscillation is $a = 6.52$.

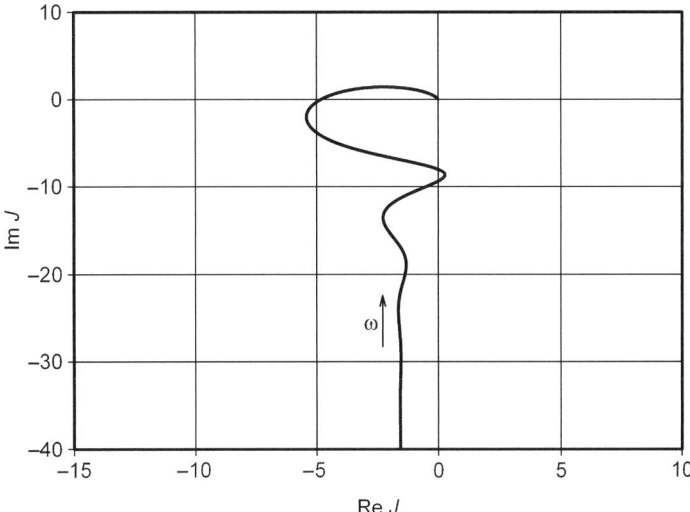

Fig. 2.12. LPRS for example of Section 2.12

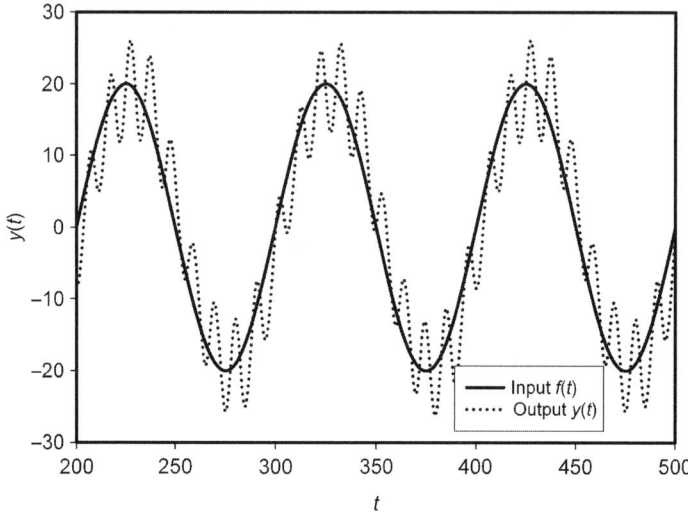

Fig. 2.13. System response to external harmonic excitation (simulations)

We assess the orbital stability of the system. The eigenvalues of the matrix Φ_0 computed per (2.34) are $\lambda_1 = -0.319$ and $\lambda_2 = 0.062$, which have magnitudes smaller than one. Therefore, the system is orbitally asymptotically stable. According to formula (2.4), we calculate the equivalent gain: $k_n = 0.103$. We assess the tracking quality of the input signal by the system using the linearized model, which is obtained via substitution of the gain k_n for the relay characteristic. Employing the methods of linear systems analysis, we obtain the component of the motion of frequency ω_{in} in the output signal. Note: The output signal also contains the self-excited periodic component $y_p(t)$ of frequency $\Omega = 0.625\text{s}^{-1}$. Therefore, the output is $y(t) = 19.8\sin(0.0628t - 0.242) + y_p(t)$, where $y_p(t)$ is the periodic component of the motion of frequency $\Omega = 0.625\text{s}^{-1}$ (note that $y_p(t)$ is not a harmonic signal). This result matches well the results obtained through simulations (Fig. 2.13).

2.13 Conclusions

The frequency domain methodology of analysis is based on the notion of the LPRS and an approach that involves substitution of the relay element with the *equivalent gain*. The LPRS comprises both the oscillatory and the transfer properties of a relay system, and succeeds even if the filtering hypothesis fails; therefore, it is a relatively universal characterization of a discontinuous control system. We prove that despite the fact that the LPRS is defined through the parameters of the periodic motion in the closed-loop system, it is actually a characterization of the linear part only. We derive three different techniques for computing the LPRS for both non-integrating and integrating linear parts, and consider certain properties of the LPRS. Finally, we demonstrate that the LPRS concept can be extended to nonlinear plants.

3
Input-output analysis of relay servo systems

3.1 Slow and fast signal propagation through a relay servo system

The LPRS method presented in the previous chapter allows one to build a linearized model of the averaged motions in a relay servo system. This is done by analyzing the system response to an external constant input. We have shown that the discontinuous system reacts to a constant input essentially like a linear system (if the averaged values of the variables are considered). Therefore, the relay nonlinearity acts as a certain equivalent gain with respect to the averaged values of the respective signals. This happens due to the "chatter smoothing" phenomenon.

It is normally assumed in describing function analysis [8, 50] that if the input is not a constant but a slowly varying signal (the meaning of the term "slow" was discussed previously), then the concept of the equivalent gain (incremental gain for DF analysis) of the relay is still applicable, and the value of the equivalent gain remains the same as in the analysis of the system response to a constant input. In the course of this analysis, the relay is replaced with the equivalent gain while the linear part remains unchanged. To be able to use the concept of the equivalent gain for non-constant input signals, we need to assume the inputs are slow compared to the oscillations. According to this assumption and the equivalent gain concept, the *external input is propagated through the relay without any lags or delays*, which results in the equivalent gain being a real number (not a complex one). There are some other approaches similar to the DF method that use the same assumption [71, 87]. However, in the setup for the equivalent gain derivation that was used above (when the input to the system was constant), the equivalent gain could only be obtained as a real number. Because we assume that the inputs are slow compared to the oscillations, we also assume the equality of the equivalent gain at a varying input to the one at a constant input.

In that respect, it is interesting to analyze if the assumption of the equivalent gain being a real number (not a complex one) is valid regardless of whether the external inputs are slow or not. Therefore, in this chapter we aim to analyze whether the hysteretic relay nonlinearity does or does not introduce any lags or delays to an input signal propagation. In other words, we aim to analyze how precise the concept of the equivalent gain is with respect to non-constant inputs in terms of the dynamics of signal propagation through the relay. For the purpose of this analysis, let us ignore the actual shape of the oscillation and consider the error signal being the sum of two sinusoidal signals of multiple frequencies

$$\sigma(t) = A_1 \sin(\omega t) + A_2 \sin\left(\frac{\omega}{M}t\right), \tag{3.1}$$

where ω and A_1 are the frequency and the amplitude of the oscillatory component, ω/M and A_2 are the frequency and the amplitude of the forced component, and M is positive integer.

Let the error signal (3.1) be applied to the hysteretic relay, so that the output of this relay is as follows,

$$u = \begin{cases} +c \text{ if } \sigma = f_0 - y \geq b \text{ or } \sigma > -b, u(t-0) = c \\ -c \text{ if } \sigma = f_0 - y \leq -b \text{ or } \sigma < b, u(t-0) = -c, \end{cases} \tag{3.2}$$

where c is the amplitude of the relay, b is the hysteresis value (half of the total hysteresis width) of the relay, and $u(t-0)$ is the control value at the time immediately preceding the current time.

Let us assume that the forced component acts on the oscillatory component in such a way that it biases the switching instants but does not change the switching pattern, so that the relay switches occur twice on the period of the oscillatory component. This happens if the following inequality holds:

$$A_1 > A_2 + b. \tag{3.3}$$

Let us analyze how these two components are propagated through the relay nonlinearity. To this end, let us write expressions for the equivalent complex gains of the oscillatory component and the forced component assessed as a ratio of the complex amplitudes of the first harmonics of the respective frequencies in the relay output to the complex amplitudes at the relay input. In essence, we write expressions for the describing functions for each of those two components. This is similar to the approach proposed by Gelb [50] and in certain respects to the approaches proposed in [56, 87],

$$N_1 = \frac{\omega}{\pi M A_1} \int_0^{2\pi M/\omega} u(t) \sin \omega t \, dt + j \frac{\omega}{\pi M A_1} \int_0^{2\pi M/\omega} u(t) \cos \omega t \, dt \tag{3.4}$$

and

3.1 Slow and fast signal propagation through a relay servo system

$$N_2 = \frac{\omega}{\pi M \, A_2} \int_0^{2\pi M/\omega} u(t) \sin\left(\frac{\omega}{M}t\right) dt \\ + j \frac{\omega}{\pi M \, A_2} \int_0^{2\pi M/\omega} u(t) \cos\left(\frac{\omega}{M}t\right) dt, \quad (3.5)$$

where $M \geq 2$ is integer. The complex gain (3.4) is investigated in [8, 50, 56]. However, the LPRS methodology does not involve the concept of the gain as expressed in (3.4), and we are particularly interested in the complex gain given in formula (3.5), especially in imaginary part.

The output of the relay is a series of square pulses with alternating values, either c or $-c$. The switches of the relay occur twice on the period $2\pi/\omega$, in accordance with the above assumption. Therefore, during the period $2\pi M/\omega$, there are $2M$ switches of the relay. Let us denote the switching times by $t_1, t_2, t_3, ..., t_{2M-1}, t_{2M}$. Then we can write for the imaginary part of N_2:

$$\begin{aligned}
\operatorname{Im} N_2 &= \tfrac{\omega}{\pi M \, A_2}\Big\{-\int_0^{t_1} c\cos\left(\tfrac{\omega}{M}t\right) dt + \int_{t_1}^{t_2} c\cos\left(\tfrac{\omega}{M}t\right) dt \\
&\quad + ... - \int_{t_{2M}}^{2\pi M/\omega} c\cos\left(\tfrac{\omega}{M}t\right) dt\Big\} \\
&= \tfrac{c\omega}{\pi M \, A_2}\tfrac{M}{\omega}\Big\{-\sin\left(\tfrac{\omega}{M}t\right)\big|_0^{t_1} + \sin\left(\tfrac{\omega}{M}t\right)\big|_{t_1}^{t_2} + ... - \sin\left(\tfrac{\omega}{M}t\right)\big|_{t_{2M}}^{2\pi M/\omega}\Big\} \\
&= \tfrac{c}{\pi A_2}\Big\{-\left(\sin\left(\tfrac{\omega}{M}t_1\right) - \sin 0\right) + \left(\sin\left(\tfrac{\omega}{M}t_2\right) - \sin\left(\tfrac{\omega}{M}t_1\right)\right) \\
&\quad + ... - \left(\sin(2\pi) - \sin\left(\tfrac{\omega}{M}t_{2M}\right)\right)\Big\} \\
&= \tfrac{2c}{\pi A_2}\Big\{-\sin\left(\tfrac{\omega}{M}t_1\right) + \sin\left(\tfrac{\omega}{M}t_2\right) - \sin\left(\tfrac{\omega}{M}t_3\right) \\
&\quad + ... - \sin\left(\tfrac{\omega}{M}t_{2M-1}\right) + \sin\left(\tfrac{\omega}{M}t_{2M}\right)\Big\}.
\end{aligned} \quad (3.6)$$

Thus, we have obtained the formula for the imaginary part of the complex gain N_2. All addends that contain odd switching instants are negative, and all addends that contain even switching instants are positive. Formula (3.6) contains switching instants that are currently unknown. We need to find these values.

At first consider the case when $A_2 = 0$ and the error signal has only one component. Obviously, the switching occurs when either $A_1 \sin \omega t = b$ or $A_1 \sin \omega t = -b$ (in addition the condition of proper direction of the relay switch must be satisfied). Denote the switching instants t_{0i}, where i is the sequential number of the switch. The switching instants can be explicitly given by the following formulas,

$$t_{0i} = \frac{1}{\omega}\arcsin\frac{b}{A_1} + \frac{\pi(i-1)}{\omega} \quad (3.7)$$

for both odd and even i.

Now we replace t_i in (3.6) with the expressions given by formula (3.7) and we find the value of $\operatorname{Im} N_2$ in the absence of the forced component at the relay input ($A_2 \longrightarrow 0$)

$$\operatorname{Im} N_2 = \frac{2c}{\pi A_2}\sum_{i=1}^{2M}(-1)^i\left[\sin\left(\frac{1}{M}\arcsin\frac{b}{A_1} + \frac{\pi(i-1)}{M}\right)\right]. \quad (3.8)$$

We notice from (3.8) that

$$\sin\left(\frac{\omega}{M}t_{M+i}\right) = -\sin\left(\frac{\omega}{M}t_i\right),$$

and if M is even, then the sum (3.8) is zero. If M is odd, then it is not obvious whether the sum becomes zero or non-zero if no forced component is applied. For that reason, assume for the purpose of our analysis that M is even.

Assume now that amplitude A_2 is small compared to A_1, so that the forced component has a biasing effect on the oscillatory component, which results in changes of the switching times. A switching time change (increment) can be estimated with the use of linear components in the Taylor series expansion of $\sigma(t)$ around the respective switching instant as follows:

$$\sigma(t) = \sigma(t_{0i}) + \dot{\sigma}(t_{0i}) \cdot (t - t_{0i}), \tag{3.9}$$

where $\sigma(t_{0i})$ is

$$\sigma(t_{0i}) = A_1 \sin \omega t_{0i} + A_2 \sin\left(\frac{\omega}{M}t_{0i}\right) \tag{3.10}$$

and $\dot{\sigma}(t_{0i})$ can be given as follows:

$$\dot{\sigma}(t_{0i}) = A_1 \omega \cos \omega t_{0i} + A_2 \frac{\omega}{M} \cos\left(\frac{\omega}{M}t_{0i}\right). \tag{3.11}$$

We write equations for the switching instants — separately for odd end even values of i. From (3.9)–(3.11) we obtain

$$A_1 \sin \omega t_{0i} + A_2 \sin\left(\frac{\omega}{M}t_{0i}\right) + \left[A_1\omega \cos \omega t_{0i} + A_2 \frac{\omega}{M} \cos\left(\frac{\omega}{M}t_{0i}\right)\right] \cdot \Delta t_1 = b \tag{3.12}$$

for odd i, where $\Delta t_i = t - t_{oi}$, and

$$A_1 \sin \omega t_{0i} + A_2 \sin\left(\frac{\omega}{M}t_{0i}\right) + \left[A_1\omega \cos \omega t_{0i} + A_2 \frac{\omega}{M} \cos\left(\frac{\omega}{M}t_{0i}\right)\right] \cdot \Delta t_1 = -b \tag{3.13}$$

for even i. We solve (3.12) and (3.13) for Δt_i taking into account expression (3.7) for t_{0i}. Equations (3.12) and (3.13) have the following solutions

$$\Delta t_i = \frac{A_2 \sin\left(\frac{\omega}{M}t_{0i}\right)}{\pm \omega\sqrt{A_1^2 - b^2} + A_2\frac{\omega}{M}\cos\left(\frac{\omega}{M}t_{0i}\right)}$$

$$= \frac{A_2 \sin\left(\frac{1}{M}\arcsin\frac{b}{A_1} + \frac{\pi(i-1)}{M}\right)}{\pm\omega\sqrt{A_1^2 - b^2} + A_2\frac{\omega}{M}\cos\left(\frac{1}{M}\arcsin\frac{b}{A_1} + \frac{\pi(i-1)}{M}\right)}, \tag{3.14}$$

where "+" applies to odd values of i, and "−" applies to even values of i.

Since $\mathrm{Im}\, N_2 = 0$ for the case of $A_2 = 0$, we can apply the superposition principle to find the first harmonic of the relay output $u(t)$. The first harmonic

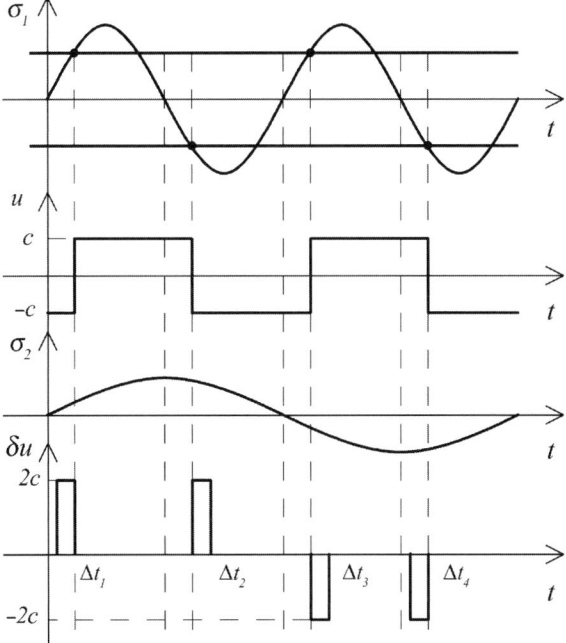

Fig. 3.1. Variation of relay output due to application of forced component

of $u(t)$ for the case of $A_2 \neq 0$ can be found as the first harmonic of the relay output variation $\delta u(t)$ (Fig. 3.1).

Since the value of A_2 is small compared with A_1, formula (3.14) can be rewritten as follows:

$$\Delta t_i \approx \frac{A_2 \sin\left(\frac{1}{m}\arcsin\frac{b}{A_1} + \frac{\pi(i-1)}{M}\right)}{(-1)^{i+1}\omega\sqrt{A_1^2 - b^2}}. \tag{3.15}$$

From (3.15), we see that for even M (that was assumed above), the following equality holds:

$$\Delta t_{M+i} = -\Delta t_i. \tag{3.16}$$

As a result, the formula for the imaginary part of the complex gain for the forced component can be rewritten as follows,

$$\operatorname{Im} N_2 = \frac{\omega}{\pi M\, A_2} \sum_{i=1}^{M} \begin{array}{l}\left[2c \cdot |\Delta t_i| \cdot \cos\left(\frac{\omega}{M}t_{0i}\right)\right. \\ \left. -2c \cdot |\Delta t_{M+i}| \cdot \cos\left(\frac{\omega}{M}t_{0M+i}\right)\right],\end{array} \tag{3.17}$$

where Δt_i is evaluated as in (3.15) and the sign is accounted for in (3.17). The following equality

$$\cos\left(\frac{\omega}{M}t_{0i}\right) = -\cos\left(\frac{\omega}{M}t_{0M+i}\right)$$

and equality (3.16), allows us to rewrite formula (3.17) as follows:

$$\text{Im } N_2 = \frac{4c\omega}{\pi M A_2} \sum_{i=1}^{M/2} |\Delta t_i| \cdot \cos\left(\frac{\omega}{M} t_{0i}\right)$$

$$= \frac{4c\omega}{\pi M A_2} \sum_{i=1}^{M/2} \left[|\Delta t_1| \cdot \cos\left(\frac{\omega}{M} t_{0i}\right) + |\Delta t_{M/2+i}| \cdot \cos\left(\frac{\omega}{M} t_{0M/2+i}\right)\right]$$

$$= \frac{4c}{\pi M \sqrt{A_1^2 - b^2}} \sum_{i=1}^{M/2} \left[\sin\left(\frac{1}{M}\arcsin\frac{b}{A_1} + \frac{\pi(i-1)}{M}\right) \cdot \cos\left(\frac{\omega}{M} t_{0i}\right)\right.$$

$$\left. - \cos\left(\frac{1}{M}\arcsin\frac{b}{A_1} + \frac{\pi(i-1)}{M}\right) \cdot \sin\left(\frac{\omega}{M} t_{0i}\right)\right]$$

$$= \frac{4c}{\pi M \sqrt{A_1^2 - b^2}} \sum_{i=1}^{M/2} \sin\left(\frac{1}{M}\arcsin\frac{b}{A_1} + \frac{\pi(i-1)}{M} - \frac{\omega}{M} t_{0i}\right) = 0. \quad (3.18)$$

We formulate the following proposition.

Proposition 3.1. *If the switching of the relay occurs two times during every period of the oscillatory component of the motion (one time from "−" to "+" and one time from "+" to "−"), so that the presence of the forced component does not change this pattern but only changes the switching instants, then the hysteretic relay does not introduce any delay (or lag) into the forced component propagation.*

This proposition is proven above when only the frequency of the oscillatory component is a multiple of the frequency of the forced component, and small amplitudes of the forced component. However, simulation suggests that Proposition 3.1 is valid for non-multiple frequencies and large amplitudes of the forced component (while (3.3) remains valid). Some simulation results are presented in Table 3.1. The value of the imaginary part of N_2 obtained

Table 3.1. Describing function for the forced component

	N_2 values for $b = 0$, $A_1 = 1$			
M	2	10	50	200
$A_2 = 0.1$	$0.638 + j0$	$0.634 + j0$	$0.645 + j0$	$0.649 + j0$
$A_2 = 0.4$	$0.636 + j0$	$0.650 + j0$	$0.650 + j0$	$0.653 + j0$
$A_2 = 0.8$	$0.636 + j0$	$0.707 + j0$	$0.706 + j0$	$0.706 + j0$
	N_2 values for $b = 0.1$			
M	2	10	50	200
$A_2 = 0.1$	$0.639 + j0$	$0.642 + j0$	$0.633 + j0$	$0.607 + j0$
$A_2 = 0.4$	$0.640 + j0$	$0.654 + j0$	$0.654 + j0$	$0.654 + j0$
$A_2 = 0.8$	$0.641 + j0$	$0.717 + j0$	$0.718 + j0$	$0.718 + j0$
	N_2 values for $b = 0.3$			
M	2	10	50	200
$A_2 = 0.1$	$0.668 + j0$	$0.668 + j0$	$0.672 + j0$	$0.659 + j0$
$A_2 = 0.4$	$0.671 + j0$	$0.688 + j0$	$0.690 + j0$	$0.693 + j0$

from the computation is negligibly small. For that reason, in Table 3.1, the imaginary part is zero.

3.2 Methodology of input-output analysis

We proved in the previous section that, subject to certain conditions imposed on the signals, the concept of the equivalent gain of the relay is valid not only for the propagation of constant and slow inputs through the relay servo system, but also for fast enough signals as long as the switching is triggered by the oscillatory component of the motion (occurs two times on the period of the oscillatory component). As a result, the input-output analysis of relay servo systems naturally follows from the LPRS method, the concept of the equivalent gain of the relay and its validity for sufficiently fast input signals.

The analysis is a three-step procedure.

At the first step, the LPRS of the relay servo system is computed (for the purpose of analysis, usually the knowledge of the whole LPRS is not necessary, only the point corresponding with the periodic solution is needed).

At the second step, the frequency of the periodic solution is found as a solution of equation (2.3).

After that, at the third step, the equivalent gain of the relay is computed using formula (2.4).

Finally, at the fourth step, the linearized model of the relay servo system for the average over the period motions is built via replacement of the relay element with the equivalent gain, and analysis of the forced signal or disturbance propagation is done with the use of this model and conventional methods of linear systems analysis.

The problems that can be solved with this methodology are the analysis of external inputs propagation through the relay servo system, which is the servo problem, and the analysis of the effect of external disturbances on the output of the system, which is the problem of analysis of disturbance attenuation in stabilization control systems. This method can also be applied to problems that are a combination of these two. In this case, the superposition principle can be used to incorporate both effects.

An example of input-output analysis is presented in the following section.

3.3 Example of forced motions analysis with the use of the LPRS

Consider the system in Fig. 3.2 with input $f = \sin 6.28t$. Input-output analysis of the relay feedback control system can be carried out as follows.

3 Input-output analysis of relay servo systems

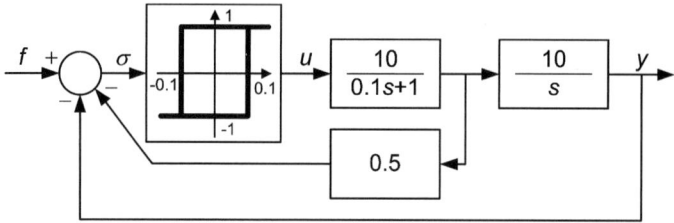

Fig. 3.2. Block diagram of the system

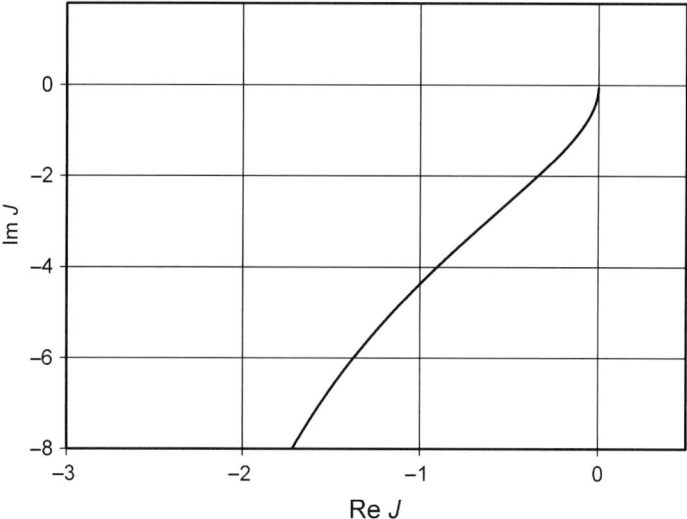

Fig. 3.3. LPRS of linear part

We expand $W_l(s)$ into the sum of $W_i(s)$,

$$W_1(s) = 100/s - 5/(0.1s+1),$$

and using the formulas of Table 2.1, we derive the formula of $J(\omega)$,

$$J(\omega) = -2.5(1 - 10\pi/\omega \operatorname{csch}(10\pi/\omega)) - j(12.5\pi^2/\omega - 1.25\pi \tanh(5\pi/\omega))$$

and plot the locus (Fig. 3.3). The intersection of the LPRS and the line parallel to the horizontal axis at the distance $\pi b/(4c) = 0.0785$ below it allows us to calculate the frequency of the oscillations and the equivalent gain of the relay k_n for the forced component of the motion. Having solved equation (2.3) for this system, we obtain $\Omega = 785.5\,\text{s}^{-1}$. According to (2.4), we calculate the gain of the relay $k_n = 714$. We substitute the gain k_n for the relay characteristic and carry out input-output analysis of the linearized system. Employing well-known methods of linear systems analysis, we obtain the forced component of the output signal. The output also contains a periodic term $z_p(t)$ with frequency $\Omega = 785.5\,\text{s}^{-1}$:

$$z(t) = 0.954 \sin(6.28t - 0.304) + z_p(t)$$

This perfectly matches the results of simulation of the system.

3.4 Conclusions

In the current chapter, we show that the LPRS theory and the equivalent gain concept can be extended to the case of external inputs being non-slow signals. Therefore, the whole approach can be applied to a much wider class of control systems. The assumption of slowness of the input signal can be removed, and the equivalent gain concept remains valid even at fast input signals as long as the relay switching is triggered by the self-excited component of the motion. These conclusions are supported theoretically and by simulations.

4

Analysis of sliding modes in the frequency domain

4.1 Introduction to sliding mode control

The sliding mode (SM) control principle is a tool to design robust controllers for nonlinear dynamic plants operating under uncertain conditions. It has become one of the most popular research areas in automatic control and has a number of industrial applications. The main advantage of the sliding mode principle is low sensitivity of the sliding mode control system to plant parameter variations and disturbances. The sliding mode principle involves a discontinuous control, which can easily be implemented as a conventional "on-off" control. This makes the controller a relatively simple device.

Usually, SM control is considered a high-speed switching feedback control because of the variable structure control type. The purpose of the switching control law is to drive the plant state trajectory onto a predetermined surface in the state space and then to maintain the state trajectory on that surface. This surface is called the *switching surface* or the *sliding surface*. The control function is designed in such a way that when the plant state trajectory is "above" the surface, the control drives the plant state "down" to the surface, and when the trajectory is "below" the surface, the control drives the plant state "up" to the surface. To realize this principle, the control is designed as a certain discontinuous function of the plant state. As a result, the plant trajectory is always driven toward the *sliding surface* by the control, so that, once intercepted, the switching control maintains the plant state trajectory on the surface. Ideally speaking, the trajectory cannot deviate from the switching surface and "slides" along it. As a result, the closed-loop system dynamics are determined by the plant state trajectory restricted to the *sliding surface*. The whole design of a SM system breaks down into two steps. The first step is to properly choose or design a sliding surface so that the closed-loop system has desired dynamics. The second step is to design a switching control that can drive the plant state trajectory onto the sliding surface and maintain it on that surface.

We do not consider the problems of SM design in this work. There are several publications devoted to the problems of SM design (see for example [38, 97, 98] and references within). Neither of the above two steps is going to be considered here. Rather we are concerned with the problem of analysis of SM systems. Therefore, we will assume that the sliding surface has been chosen and the switching control has been designed. The objective of this chapter is to present a methodology for the analysis of SM systems based on the LPRS approach.

The SM described above, with the state trajectory strictly located on the sliding surface, is usually referred to as an ideal SM. In the ideal SM, the switching of the discontinuous control occurs with infinite frequency. Yet the real SM is realized as high but not infinite frequency oscillations. This phenomenon is usually referred to as chattering, which remains the main drawback of sliding mode control [105]. Chattering is caused by the presence of an actuator/sensor and/or switching imperfections. The presence of an actuator (or sensor or switching imperfections) results in the convergence of the transient process to the steady state, which is not an equilibrium point but a periodic motion, and also results in the existence of oscillations in the transient process. There also exist applications where the control must be implemented as a pulse-width modulated signal and the chattering is a normal operating mode of the system. In both cases, the analysis of chattering is an important theoretical and especially practical problem. Another manifestation of the real sliding mode in comparison to the ideal one is the distinction between the slow motions (the system output in particular) in a real sliding mode control system from the slow motions in the reduced order system. It not only concerns the character of the transient process but also the steady-state value. The average motions in a real sliding mode may converge not to the origin but to a different point, due to external disturbances [26]. Although research efforts have been focused mostly on the problems related to ideal sliding, there are a number of works devoted to the analysis of real sliding modes [33, 45, 48, 85, 88, 97] as well as to the design methods aimed at chattering reduction [11, 27, 49, 65, 67, 89, 92]. The problem of the construction of a more precise model for the slow motions than that of the reduced order system has been addressed only for the relay type of control with the use of the describing function method [93, 108]. This problem may seem to be of less importance, as there is a fairly good approximation of the slow motions, which is the solution of the reduced-order system. However, the reduced-order model cannot address some important phenomena in SM control systems and, in our opinion, the indicated problem deserves further attention, as its solution can be useful for real-world applications. The use of the equivalent gain and linearization based on that concept can also transfer the original problem into the framework of classic feedback design with applications to sliding mode control.

If the discontinuous control in a SM system is implemented as a relay control, the LPRS method presented previously can be used for analysis of chattering and for solving the input-output problem in a SM control system.

However, very often the switching control is designed as a discontinuous, but not a relay-type, control. In this case, the LPRS approach cannot be applied to the analysis of such a system directly. Yet the distinction of the control algorithm from the relay control is not a limitation if an ideal SM is analyzed, since the dynamics of the system are reduced in exactly the same way for all types of control because the trajectory is restricted to the sliding surface. Therefore, the closed-loop system dynamics do not depend on the type of the switching control so long as the ideal sliding occurs. This is not the case if we consider the real SM manifested as chattering. Here, the motions are not restricted to the sliding surface. The motion occurs in a certain vicinity of the sliding surface, and the motion depends on the type of discontinuous control [18]. As a result, the closed-loop system dynamics depend on the type of discontinuous control. Most commonly, the following types of discontinuous control are used to generate a SM. We will refer to them as control algorithms (relay control is included in this list, too):

- relays with constant output amplitudes (relay feedback systems),
- relays with state-dependent output amplitudes,
- linear state feedback control with switched gains,
- a sum of the discontinuous and equivalent control, and
- a combination of the above.

We solve the problem of analysis of a SM system by obtaining a relay feedback system equivalent to the original SM system in a certain sense, and analyzing the *equivalent relay system* with the use of the LPRS methodology. We make the following assumptions in this problem. The real sliding mode exhibits two types of motions: the fast motions associated with the motion across the sliding surface and the slow motions associated with the motion along the sliding surface. Also, the shape of the control signal in the time domain is close to a square shape regardless of the realization of the control function.

At first, we consider the problem of approximating the original control function by the relay function in a vicinity of the sliding surface. After that, we solve the problem of decomposing the original system into fast and slow motion dynamics. Then we will apply the LPRS method to the obtained structure and develop the analysis methodology. We also obtain a description of ideal sliding as a limiting case of the non-reduced order model. Finally, we consider an example of analysis of a SM system.

4.2 Representation of a sliding mode system via the equivalent relay system

Consider a single-input single-output linear time-invariant plant:

$$\dot{\mathbf{x}} = \mathbf{A}\mathbf{x} + \mathbf{B}u \tag{4.1}$$

where $\mathbf{x} \in R^n$ is a state vector, $u \in R^1$ is the control, $\mathbf{A}$ is an $n \times n$ matrix, and $\mathbf{B}$ is an $n \times 1$ matrix. We assume that the model (4.1) includes both principal and parasitic dynamics, which are discussed below. The presence of parasitic dynamics results in chattering, so that the control signal is a switching control of finite frequency. The scalar control is a function of the state vector

$$u(\mathbf{x}) = \begin{cases} u^+(\mathbf{x}) \text{ if } \sigma(\mathbf{x}) > 0 \\ u^-(\mathbf{x}) \text{ if } \sigma(\mathbf{x}) < 0 \end{cases} \quad (4.2)$$

where $u^+(\mathbf{x})$ and $u^-(\mathbf{x})$ are two different control functions, and the sliding surface is a hyperplane given by the following equality,

$$\sigma = \mathbf{S}\mathbf{x} = 0 \quad (4.3)$$

where $\mathbf{S}$ is a row matrix of coefficients.

The control $u(\mathbf{x})$ is designed in such a way that the tangent vectors of the state trajectory point toward the sliding surface. In the time-invariant case, it is given as a function of the state vector only. The sliding surface divides the state space into two subspaces. Normally, the control $u(\mathbf{x})$ is defined everywhere except for the sliding surface. We call the controls $u^+(\mathbf{x})$ and $u^-(\mathbf{x})$ the *subspace controls* "above" the surface and "below" the surface, respectively. We proceed from the assumption that the *subspace controls* "above" the surface $u^+(\mathbf{x})$ and "below" the surface $u^-(\mathbf{x})$ are defined at every point of the state space, including the sliding surface, and are differentiable functions of $\mathbf{x}$ within the range of sign constancy of the state variables $x_i, i = \overline{1, n}$. Note that the switching control $u(\mathbf{x})$ may not be defined on the sliding surface due to the selection logic given in (4.2). Most of the control functions used as *subspace controls* satisfy this condition.

Assume that *there exists a $\triangle$-neighborhood of the sliding surface such that once the trajectory enters the $\triangle$-neighborhood, it stays within this neighborhood for all subsequent time.* For any type of control listed above, we assume that in the $\triangle$-neighborhood of the sliding surface, each of the subspace controls $(u^+(\mathbf{x})$ and $u^-(\mathbf{x}))$, as functions of $\mathbf{x}$, can be expanded into Taylor series. Namely, the control functions can be expanded into Taylor series at every point of the sliding surface, as a series of the deviations from the surface.

The term "deviation from the sliding surface" requires some elaboration. We do not consider the system dynamics at this time; we try, instead, to construct a suitable approximation of the control $u(\mathbf{x})$ as a function of the state vector $\mathbf{x}$ in the state-control space. For that reason, we can set the states $x_1, x_2, ..., x_{n-1}$, equal to the respective coordinates of a point on the sliding surface $x_{S1}, x_{S2}, ..., x_{Sn-1}$, and assume that x_n is the only variable that undergoes a deviation from the sliding surface. Under this supposition, the condition that the location of a point of the trajectory be within the $\triangle$-neighborhood is

4.2 Representation of a sliding mode system via the equivalent relay system

$$\left|\frac{\sigma(\mathbf{x})}{s_n}\right| < \triangle \quad (4.4)$$

where s_n is the n-th coefficient of the matrix $\mathbf{S}$. Let us denote the control "just above" the sliding surface (when $\sigma(\mathbf{x}) = 0+$) as $u^+(\mathbf{x}_s)$ and the control "just below" the sliding surface (when $\sigma(\mathbf{x}) = 0-$) as $u^-(\mathbf{x}_s)$. In fact, those control functions are the *subspace controls* on the sliding surface, so that $u^+(\mathbf{x}_s) := \{u^+(\mathbf{x})|\mathbf{S}\mathbf{x} = 0\}$ and $u^-(\mathbf{x}_s) := \{u^-(\mathbf{x})|\mathbf{S}\mathbf{x} = 0\}$. Then at every point of the $\triangle$-*neighborhood*, the *subspace controls* can be approximated by the series expansion

$$\begin{aligned} u^+(\mathbf{x}) &= u^+(\mathbf{x}_s) + \frac{\partial u^+}{\partial x_n} \triangle x_n + \frac{\partial^2 u^+}{\partial x_n^2} \triangle x_n^2 + \ldots \\ u^-(\mathbf{x}) &= u^-(\mathbf{x}_s) + \frac{\partial u^-}{\partial x_n} \triangle x_n + \frac{\partial^2 u^-}{\partial x_n^2} \triangle x_n^2 + \ldots \end{aligned} \quad (4.5)$$

where the derivatives are taken at the point $\mathbf{x} = \mathbf{x}_s$, and $\triangle x_n$ is the deviation of x_n from the sliding surface and is given by

$$\triangle x_n = \frac{\sigma(\mathbf{x})}{s_n}. \quad (4.6)$$

Each of the *subspace control* functions has now two terms: a term that depends on the variables $x_1, x_2, \ldots, x_{n-1}$ and a term that depends only on the variable x_n. We shall call the first term "the relay term" and the second one "the continuous term." It is important for future analysis that the relay term does not depend on the variable x_n. We write the expression for the overall switching control and transform the obtained approximate control function into the following form:

$$u = \psi_1(\mathbf{x}^*)\text{sign}\,\sigma + \psi_2(\mathbf{x}^*) + \rho(\sigma) \quad (4.7)$$

where $\psi_1(\mathbf{x}^*)\text{sign}\,\sigma + \psi_2(\mathbf{x}^*)$ is the relay term and $\rho(\sigma)$ is the continuous term, $\mathbf{x}^*$ is a subset of $\mathbf{x}$, and $\mathbf{x}^* \in R^{n-1}$ is the vector $\mathbf{x}$ without element x_n,

$$\psi_1(\mathbf{x}) = \frac{u^+(\mathbf{x}_s) - u^-(\mathbf{x}_s)}{2} \quad (4.8)$$

$$\psi_2(\mathbf{x}) = \frac{u^+(\mathbf{x}_s) + u^-(\mathbf{x}_s)}{2} \quad (4.9)$$

$$\rho(\sigma) = \begin{cases} \frac{\partial u^+}{\partial x_n}\frac{\sigma}{s_n} + \frac{\partial^2 u^+}{\partial x_n^2}\frac{\sigma^2}{s_n^2} + \ldots \text{if } \sigma \geq 0 \\ \frac{\partial u^-}{\partial x_n}\frac{\sigma}{s_n} + \frac{\partial^2 u^-}{\partial x_n^2}\frac{\sigma^2}{s_n^2} + \ldots \text{if } \sigma < 0. \end{cases} \quad (4.10)$$

We note that the functions $\psi_1(\mathbf{x}^*)$ and $\psi_2(\mathbf{x}^*)$ are continuous, since the *subspace controls* are continuous functions of their arguments. The function $\rho(\sigma)$ is continuous but does not necessarily have a continuous first derivative, and also $\rho(0) = 0$. The function $\psi_1(\mathbf{x}^*)$ is always positive, since it is the difference between the corresponding higher and lower values of the *subspace*

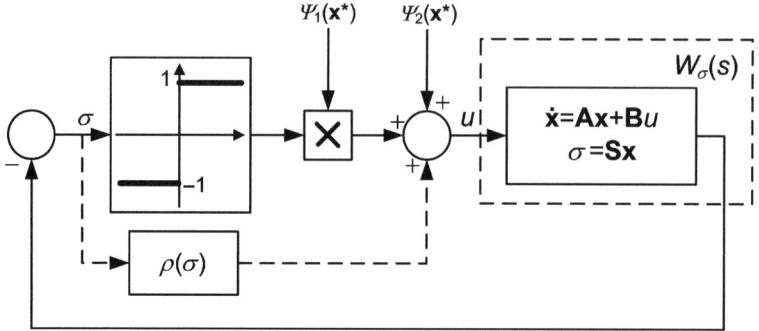

Fig. 4.1. Equivalent relay system

controls (see (4.8) above). A reasonable approach to approximation of $u^+(\mathbf{x})$ and $u^-(\mathbf{x})$ is to limit the series (4.5) to the linear terms. In that case, the function $\rho(\sigma)$ is a linear gain or a combination of two gains which are different for positive and negative argument values. As examples of the design of SM control systems show, in many cases the function $\rho(\sigma)$ may not be taken into account at all. Disregarding this component often just slightly reduces the accuracy of analysis. Another justification of such an approach is that the purpose of the current analysis is not to build a very precise model of the slow and fast motions in a sliding mode control system but to build a model that is more accurate than the commonly used reduced-order model (considered below) on the one hand and still takes account of all essential phenomena in the system. Very high precision can always be achieved by integration of the original equations of the system (containing the model of imperfections). This approach results in the following block diagram representation, which is further referred to as the equivalent relay system (the link through the function $\rho(\sigma)$ is shown as a dashed line because it may or may not be used, depending on the required accuracy).

The model in Fig. 4.1 denoted by the transfer function $W_\sigma(s)$ refers to the combined principal and parasitic dynamics. We have thus completed the regularization and approximation of the control function. The brief example below illustrates this methodology.

Example 4.1. Let the plant dynamics be given by (4.1) and the sliding surface by

$$\sigma = x_1 + 1.5\, x_2 + x_3.$$

This corresponds to the desired reduced-order dynamics with characteristic polynomial $P(\lambda) = \lambda^2 + 1.5\,\lambda + 1$. Also, let the control be designed as a linear state feedback control with switched gains:

$$u(\mathbf{x}) = k_1(x_1)x_1 + k_2(x_2)x_2 + k_3(x_3)x_3,$$

where

$$k_i(x_i) = \begin{cases} \alpha_i \text{ if } x_i > 0 \\ \beta_i \text{ if } x_i < 0 \end{cases}, \quad i = \overline{1,3}$$

Let the trajectory be fully located in the octant given by $x_1 > 0$, $x_2 < 0$, $x_3 > 0$. Then the *subspace controls* in that octant are

$$u^+(\mathbf{x}) = \alpha_1 x_1 + \beta_2 x_2 + \alpha_3 x_3,$$
$$u^-(\mathbf{x}) = \beta_1 x_1 + \alpha_2 x_2 + \beta_3 x_3.$$

On the sliding surface $\sigma = x_1 + 1.5\, x_2 + x_3 = 0$, the *subspace controls* are (considering that $x_3 = -x_1 - 1.5\, x_2$):

$$u_s^+(\mathbf{x}_s) = (\alpha_1 - \alpha_3)x_1 + (\beta_2 - 1.5\alpha_3)x_2$$
$$u_s^-(\mathbf{x}_s) = (\beta_1 - \beta_3)x_1 + (\alpha_2 - 1.5\beta_3)x_2.$$

The derivatives of the *subspace controls*, necessary for calculation of the continuous term, (4.10) are $\partial u^+/\partial x_3 = \alpha_3$ and $\partial u^-/\partial x_3 = \beta_3$. Therefore, the amplitude of the symmetric relay control is

$$\psi_1(\mathbf{x}^*) = 0.5((\alpha_1 - \alpha_3 - \beta_1 + \beta_3)x_1 + (\beta_2 - 1.5\alpha_3 - \alpha_2 + 1.5\beta_3)x_2).$$

and the bias of the relay control is

$$\psi_2(\mathbf{x}^*) = 0.5((\alpha_1 - \alpha_3 + \beta_1 - \beta_3)x_1 + (\beta_2 - 1.5\alpha_3 + \alpha_2 - 1.5\beta_3)x_2).$$

The continuous term of the control function is $\rho(\mathbf{x}) = \alpha_3 \sigma$ if $\sigma \geq 0$ and $\rho(\mathbf{x}) = \beta_3 \sigma$ if $\sigma < 0$.

Therefore, we obtain an approximation suitable for the following analysis.

4.3 Analysis of motions in the equivalent relay system

Now we consider analysis of motions in the equivalent relay system (Fig. 4.1). Obviously, there are two simultaneous motions in this system: the fast motion, denoted by $\mathbf{x}_f(t)$, and the slow motion, denoted by $\mathbf{x}_0(t)$:

$$\mathbf{x}(t) = \mathbf{x}_0(t) + \mathbf{x}_f(t). \tag{4.11}$$

The fast motion, manifested as high-frequency oscillation (chattering), occurs due to the existence of the relay component of the control, and the slow motion occurs due to the existence of the non-zero average control value (averaged over the period of the fast motion) and non-zero initial conditions. The notion of averaged control is similar to the notion of equivalent control but not identical to it, as the model of the slow motions is not the reduced-order model. Usually, the frequency of the fast motion is much higher than the frequency of the slow motion. This allows for the following two assumptions:

74   4 Analysis of SM in the frequency domain

Assume that the slow component $\mathbf{x}_0(t)$ *remains constant during the period of the fast motions; and also assume that* $\psi_1(\mathbf{x}^*) = \psi_1(\mathbf{x}_0^*)$ *and* $\psi_2(\mathbf{x}^*) = \psi_2(\mathbf{x}_0^*)$, *where* $\mathbf{x}_0^*$ *is the slow component of the truncated state vector (vector* $\mathbf{x}$ *without element* x_n). The first assumption is the same as the one made in the analysis of relay feedback systems with the use of the LPRS. The second assumption is the assumption of independence of the amplitude and the bias of the relay control function from the fast component of the motion. In essence, we assume that the amplitude bias of the discontinuous component of the control $\psi_1(\mathbf{x}^*)\text{sign } \sigma + \psi_2(\mathbf{x}^*)$ does not change during the period of one oscillation even if $\mathbf{x}^*$ changes; on one period of oscillation, the control value changes only due to the continuous term change $\rho(\sigma)$. The closeness of $\mathbf{x}^*$ and $\mathbf{x}_0^*$ and the validity of the second assumption are illustrated by the following example. Let $x_n(t)$ be the sum of two sinusoidal signals, $x_n(t) = \sin(\omega t) + \sin(h\omega t)$, where the first component is slow motion and the second component is fast motion. We suppose that h is a large number. Then $x_{n-1}(t)$, being the integral of $x_n(t)$, is given by $x_{n-1}(t) = -\omega^{-1}\cos(\omega t) - h^{-1}\omega^{-1}\cos(h\omega t)$, and therefore, the amplitude of the fast component is decreased by a factor of h compared to the slow component, and so on, inductively down to the variable $x_1(t)$. The difference between $\mathbf{x}^*$ and $\mathbf{x}_0^*$ lies only in the existence of the high-frequency component in $\mathbf{x}^*$. Yet we have shown that if h is a large number, $\mathbf{x}^*$ and $\mathbf{x}_0^*$ are close, because the amplitude of the high-frequency component is small. Therefore, once this property is applied to the estimation of the functions $\psi_1(\mathbf{x})$ and $\psi_2(\mathbf{x})$, we obtain the "quasi-statical" model of the change in the discontinuous component of the control $\psi_1(\mathbf{x}^*)\text{sign } \sigma + \psi_2(\mathbf{x}^*)$ from one period of oscillations to another. Moreover, we showed that in the relay feedback system, both the frequency of the fast motions and the equivalent gain of the relay do not depend at all on the value of the relay amplitude. This is also applicable to the equivalent relay system under present consideration, which also justifies the "quasi-statical" model of the control amplitude change.

Therefore, the above assumptions allow us to apply the "frozen parameters" principle to the analysis of fast motions and the principle of "chatter smoothing" to the analysis of slow motions, with the replacement of the relay nonlinearity with the equivalent gain. As a result, the models of fast and slow motions can be represented by the diagram in Fig. 4.2. In Fig. 4.2, the block containing the function $\rho(\sigma)$ is transposed to the feedback around the plant, which obviously is equivalent to the original structure (Fig. 4.1).

We now have two models for the fast (upper) and for the slow (lower) motion analysis. Those models are not independent but interact with each other through a set of parameters that are a result of each system's solution.

Under the first assumption, obtain the solution of the "fast" system. We shall further consider only the piecewise-linear form of the function $\rho(\sigma)$ (only linear terms of (4.10)). Viewing the "fast" system as a system with a limit cycle, we obtain the Poincaré return map for it. The model of the fast motions can be written as

$$\dot{\mathbf{x}} = \mathbf{A}\mathbf{x} + \mathbf{B}u \tag{4.12}$$

4.3 Analysis of motions in the equivalent relay system

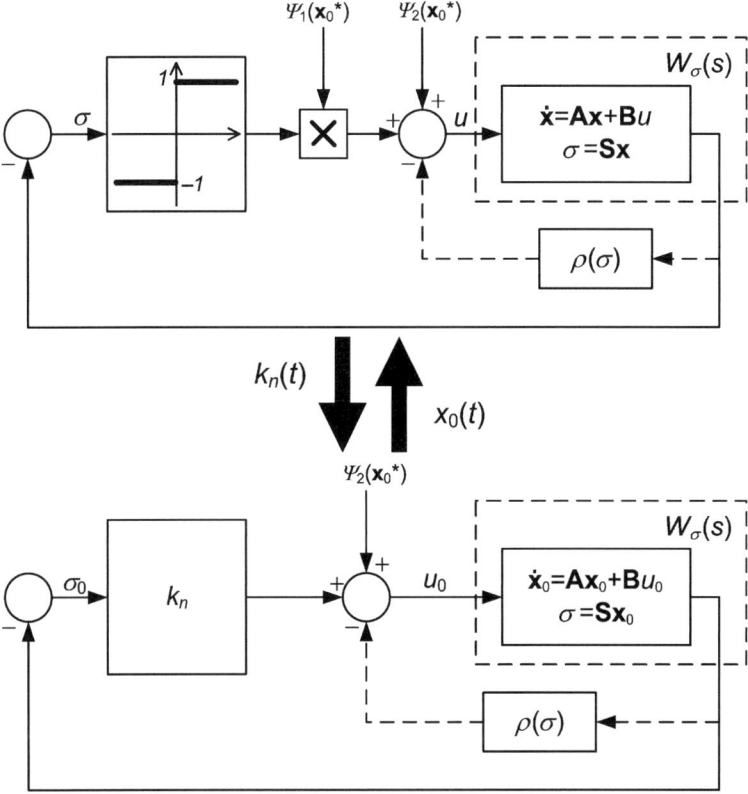

Fig. 4.2. Model of fast and slow motions

where $\mathbf{x} \in R^n, u \in R^1, \mathbf{A} \in R^{n \times n}, \mathbf{B} \in R^{n \times 1}$, with a scalar control given by

$$u(\mathbf{x}) = \psi_1 \text{sign}(\sigma) + \psi_2 + \rho(\sigma) \tag{4.13}$$

where ψ_1 and ψ_2 are constant values, $\rho = \rho_1 \sigma$ if $\sigma \geq 0$ or $\rho = \rho_2 \sigma$ if $\sigma < 0$ (ρ_1 and ρ_2 are constants), and the sliding surface is a hyperplane given by

$$\sigma = \mathbf{Sx} \tag{4.14}$$

where $\mathbf{S}$ is a $1 \times l$ matrix of coefficients. We take account of the link through the function $\rho(\sigma)$ by making adjustments to the matrix $\mathbf{A}$. Because $\rho(\sigma)$ is piecewise linear and $\sigma(t)$ is a signal that causes both switches of the control and switches of $\rho(\sigma)$ between $\rho = \rho_1 \sigma$ and $\rho = \rho_2 \sigma$, we consider the following switched plant model. Denote by $\mathbf{A}_1$ the matrix which is constructed from matrix $\mathbf{A}$ by replacing element a_{11} with element $(a_{11} - \rho_1)$, and $\mathbf{A}_2$ the matrix constructed from $\mathbf{A}$ by replacing element a_{11} with element $(a_{11} - \rho_2)$. (Note: element a_{11} is the outer gain a_1 of the control canonical form of the plant description with a negative sign.) By doing this, we write two separate sets of equations of the plant, one for the positive control and one for the negative control, to find the periodic solution.

A common way to find a periodic solution is to use a Poincaré map. Because the control switches are unequally spaced and the oscillations are not symmetric, we must consider a Poincaré return map. Suppose an asymmetric periodic process with period T exists in the system. Then, by recalling the solution for the constant control $u = \pm 1$,

$$\mathbf{x}(t) = e^{\mathbf{A}t}\mathbf{x}(0) + \mathbf{A}^{-1}(e^{\mathbf{A}t} - \mathbf{I})\mathbf{B}u$$

the periodic solution of system (4.12), (4.13) can be written as

$$\eta = e^{\mathbf{A}_1\theta_1}\zeta + \mathbf{A}_1^{-1}(e^{\mathbf{A}_1\theta_1} - \mathbf{I})\mathbf{B}\psi_1, \tag{4.15}$$

$$\zeta = e^{\mathbf{A}_2\theta_2}\eta - \mathbf{A}_2^{-1}(e^{\mathbf{A}_2\theta_2} - \mathbf{I})\mathbf{B}\psi_1, \tag{4.16}$$

where $\zeta = \mathbf{x}(0) = \mathbf{x}(T)$, $\eta = \mathbf{x}(\theta_1)$ for the periodic solution, and θ_1, θ_2 are the positive and the negative pulse duration of the periodic control $u(t)$. Formulas (4.15) and (4.16) are a Poincaré return map for the system (sequential numbers of switches are not shown). The effect of ψ_2 is not considered in (4.15), (4.16). It is more convenient to add this constant value to the sliding variable $\sigma(t)$. Solving equations (4.15),(4.16) gives

$$\zeta = (1 - e^{\mathbf{A}_2\theta_2}e^{\mathbf{A}_1\theta_1})^{-1}[e^{\mathbf{A}_2\theta_2}\mathbf{A}_1^{-1}(e^{\mathbf{A}_1\theta_1} - \mathbf{I}) - \mathbf{A}_2^{-1}(e^{\mathbf{A}_2\theta_2} - \mathbf{I})]\mathbf{B}\psi_1 \tag{4.17}$$

$$\eta = (1 - e^{\mathbf{A}_1\theta_1}e^{\mathbf{A}_2\theta_2})^{-1}[-e^{\mathbf{A}_1\theta_1}\mathbf{A}_2^{-1}(e^{\mathbf{A}_2\theta_2} - \mathbf{I}) + \mathbf{A}_1^{-1}(e^{\mathbf{A}_1\theta_1} - \mathbf{I})]\mathbf{B}\psi_1 \tag{4.18}$$

and the switching conditions are

$$\sigma(0) = \mathbf{S}(\zeta - \mathbf{A}_2^{-1}\mathbf{B}\psi_2) = 0 \tag{4.19}$$

$$\sigma(\theta_1) = \mathbf{S}(\eta - \mathbf{A}_1^{-1}\mathbf{B}\psi_2) = 0. \tag{4.20}$$

In (4.19), (4.20), the effect of ψ_2 is added.

Hence, we obtain a set of two equations (4.19), (4.20) (with substitution of (4.17) and (4.18)), from which the positive and negative control pulse length θ_1, θ_2 can be found. From this, we can obtain the frequency of the oscillations and the equivalent gain of the relay as a quotient of constant terms of the relay output and relay input. The slow motions can be analyzed as the motions in the system Fig. 4.2 (lower part). The transfer properties of the system can be analyzed as a response of the system output to the constant input ψ_2 (or another external input or disturbance).

Although there is no fundamental difficulty in the solution of system (4.17)–(4.20), this system is not simple enough to serve as a basis for qualitative analysis and design conclusions. The limiting case of equations (4.17), (4.18), when the asymmetry of the control tends to zero, ($\theta_1 \rightarrow T/2$, $\theta_2 \rightarrow T/2$, $\psi_2 \rightarrow 0$ with T being the period) and $\mathbf{A}_1 = \mathbf{A}_2 = \mathbf{A}$, provides more opportunities in this respect without a significant loss of accuracy. We, therefore, have arrived at the LPRS method, which, as illustrated in Chapter 2, considers asymmetric oscillations corresponding to unequally spaced relay switching with infinitesimally small asymmetry. Despite the fact that the LPRS method described in the preceding chapters and the analysis of the real sliding phenomena is fundamentally no different than analysis of relay servo systems, we consider the LPRS analysis of SM control systems, because this analysis deals with some specific problems of SM control that are not typical of conventional relay servo systems.

4.4 The chattering phenomenon and its LPRS analysis

We have shown that the SM control system is essentially a relay feedback system even if the control algorithm is not a relay function. This is very important and allows us to carry out analysis of SM systems as relay feedback systems without taking into account the actual SM control algorithm.

The real SM is always realized as high but finite frequency oscillations. This phenomenon is known as chattering [22, 45, 46, 97, 98, 105]. Chattering is caused by the presence of an actuator/sensor and/or switching imperfections. The presence of an actuator results in the convergence of the transient process to a steady periodic motion instead of an equilibrium point. Oscillations also exist in the transient process. In Section 4.2, when deriving the equations of the equivalent relay system, we intentionally mentioned the existence of parasitic dynamics in the system model. Without parasitic dynamics, an ideal SM would occur, and the system transformation into an equivalent relay form would be unnecessary, as chattering would not occur. The subject of this chapter is the analysis of real SM phenomena, which manifest themselves as chattering and non-ideal closed-loop performance (non-ideal tracking of external signals or non-ideal disturbance rejection). From now on, with the methodology of transforming any SM system into an equivalent relay system available, we consider only one type of SM algorithm: the relay feedback algorithm (system). We find conditions for the existence of the ideal SM and chattering by applying the LPRS method.

Nevertheless, we note the difference between the treatment of the ideal SM in the approach based on Lyapunov stability and in the LPRS approach. If we consider the relay feedback system in Fig. 1.3 and the Poincaré map given by (2.5) and (2.6) (we assume a symmetric map), the whole idea of the existence of ideal SM is based upon the absence of a fixed point of this Poincaré map (Fig. 4.3.a). A different treatment of ideal SM arises through the LPRS approach. The LPRS is a frequency-domain method, which implies that a periodic motion exists in the system. As per the LPRS approach, ideal SM occurs when the frequency of this periodic motion becomes infinite. This situation is a limiting case of the system with a limit cycle of finite frequency at one value of a certain variable parameter (the relay hysteresis, for example), which becomes the limit cycle of infinite frequency if this parameter tends to

78   4 Analysis of SM in the frequency domain

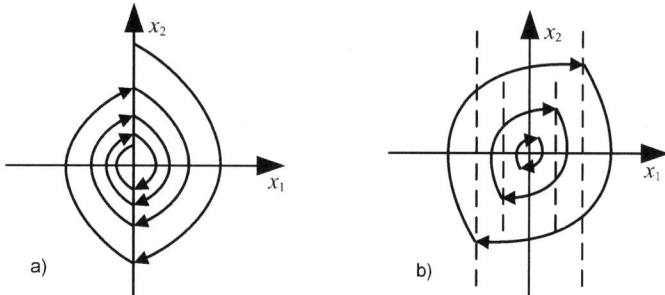

Fig. 4.3. Poincaré maps: (a) in system having ideal SM and (b) in system having a limit cycle (with a few different values of hysteresis of relay)

a certain limiting value (the hysteresis tending to zero, for example). This approach is illustrated in Fig. 4.3.b. Chattering, on the other hand, can be considered as a limit cycle or as the existence of a fixed point of the Poincaré map (2.5), (2.6).

SM control is designed in a certain specified way that, despite the fact that a SM system is a relay feedback system, it is a special type of relay feedback system, which necessitates some special treatment of a SM system.

We now analyze the SM control system consisting of a plant and an actuator and represent the equations of a variable structure SM system (4.1)–(4.3) as a relay type SM control system in the following format,

$$\dot{\mathbf{x}}_p = \mathbf{A}_p \mathbf{x}_p + \mathbf{B}_p u_a \tag{4.21}$$

where $\mathbf{x}_p \in R^l$ is the state vector of the plant, $\mathbf{x}_p \subset \mathbf{x}$, $u_a \in R^1$ is the control applied to the plant (the actuator output), and $\mathbf{A}_p \in R^{l \times l}$, $\mathbf{B}_p \in R^{l \times 1}$. The scalar control produced by the SM control algorithm is

$$u(\mathbf{x}_p) = \begin{cases} c & \text{if } \sigma(\mathbf{x}_p) > 0 \\ -c & \text{if } \sigma(\mathbf{x}_p) < 0 \end{cases} \tag{4.22}$$

and the sliding surface is given by

$$\sigma = \mathbf{S}_p \mathbf{x}_p = 0 \tag{4.23}$$

where $\mathbf{S}_p$ is an $1 \times l$ row matrix of coefficients, $\mathbf{S}_p \subset \mathbf{S}$.

The dynamics given by formulas (4.21)–(4.23) are called the *principal dynamics* of the SM system. The principal dynamics, therefore, include the dynamics of the plant and the equation of the sliding surface. We call variable σ the *sliding variable*. We can see from (4.22) that the input to the relay is the sliding variable. Therefore, the SM system is a relay feedback system with respect to the sliding variable, which makes it a special kind of relay feedback system. If we analyze the principal dynamics with u as the input into and σ

4.4 The chattering phenomenon and its LPRS analysis

as the output out of these dynamics (this is equivalent to the plant dynamics in the conventional sense), then we notice that the principal dynamics are always of *relative degree one* (noting that $\mathbf{S}_p \equiv \mathbf{S}$ and assuming that n-th element of matrix $\mathbf{S}$ is non-zero). This is important for understanding why, if we analyze only the principal dynamics of the SM system, we always obtain the ideal SM (the relationship between relative degree of the linear part and the possibility of ideal SM to occur is discussed below). Therefore, to obtain the parameters of the mode that would occur in a real SM control system, the dynamics of the actuator, sensor, and non-ideality of the controller need to be accounted for. We call these additional dynamics the *parasitic dynamics*. We show below that chattering in a SM system exists due to the presence of these parasitic dynamics. We model parasitic dynamics by the transfer function $W_a(s)$ attributed to the actuator dynamics, which is convenient because the principal and parasitic dynamics are connected in series:

$$\frac{u_a(s)}{u(s)} = W_a(s).$$

With the plant, sliding surface, and parasitic dynamics models available, analysis of chattering is no different than analysis of periodic motions in a conventional relay feedback system. The combined dynamics of the plant, sliding surface, and parasitic dynamics are treated as a linear part of the relay system. The LPRS for this linear part must be computed at various frequencies, and the LPRS plot can be drawn on the complex plane. Once the LPRS is computed, the frequency Ω of chattering can be determined from (2.3). Equation (2.3) with $J(\omega)$ in the format with state-space equation matrices (2.12) or the transfer function infinite-series format (2.19) provides a more precise model of the chattering than, for example, [33], in which chattering is studied in the frequency domain, too (chattering in [33] is analyzed as the result of a hysteresis of the relay, and the problem is reduced to the analysis of oscillations in a system with a hysteretic relay and an integrator).

The LPRS, being a function of the frequency, contains all possible periodic solutions for a given linear part, including the solution of infinite frequency corresponding to the ideal SM. Because a periodic solution is a point of intersection between the LPRS and the real axis, the location of the high-frequency segment of the LPRS can be very informative with respect to whether ideal SM or chattering will occur in the system. If, for example, the high-frequency segment of the LPRS is located in the upper half-plane — and, therefore, the LPRS must have an intersection with the real axis at a finite frequency — then chattering normally occurs (there may be situations when both finite and infinite periodic solutions can occur).

We analyze the location of the high-frequency segment of the LPRS for arbitrary-order linear dynamics. Let the transfer function $W_l(s) = W_a(s) \cdot W_p(s)$ for the linear part (that includes the plant, the sliding surface, and possible parasitic dynamics), where $W_p(s)$ is the plant-sliding surface transfer function, determined by $W_p(s) = \mathbf{S}_p(\mathbf{I}s - \mathbf{A}_p)^{-1}\mathbf{B}_p$, be given as a quotient of two polynomials of degrees m and n,

$$W_l(s) = \frac{B_m(s)}{A_n(s)} = \frac{b_m s^m + b_{m-1} s^{m-1} + \ldots + b_0}{a_n s^n + a_{n-1} s^{n-1} + \ldots + a_0}. \quad (4.24)$$

The relative degree of the transfer function $W_l(s)$ is $(n-m)$. Thus the following statement holds (we give it without proof as it is rather straightforward; moreover it is a reflection of the well-known fact that the Nyquist plot has the real or imaginary axis as an asymptote).

Lemma 4.2. *If the function $W_l(s)$ is strictly proper $(n > m)$, then there exists ω^* corresponding to any given $\varepsilon > 0$ such that for every $\omega \geq \omega^*$:*

$$\left| \text{Re} W_l(j\omega) - \text{Re} \frac{b_m}{a_n (j\omega)^{n-m}} \right| \leq \varepsilon \left(\frac{\omega^*}{\omega} \right)^{n-m} \quad (4.25)$$

$$\left| \text{Im} W_l(j\omega) - \text{Im} \frac{b_m}{a_n (j\omega)^{n-m}} \right| \leq \varepsilon \left(\frac{\omega^*}{\omega} \right)^{n-m}. \quad (4.26)$$

This lemma is given without proof here (for an idea of the proof, see the proofs of other statements below). This lemma simply means that at frequency $\omega \geq \omega^*$, the following equality holds:

$$W_l(s) \approx \frac{b_m}{a_n s^{n-m}}.$$

Lemma 4.3. *(monotonicity of the high-frequency segment of the LPRS). If $\text{Re} W_l(j\omega)$ and $\text{Im} W_l(j\omega)$ are monotone functions of frequency ω and $|\text{Re} W_l(j\omega)|$ and $|\text{Im} W_l(j\omega)|$ are decreasing functions of frequency ω for every $\omega \geq \omega^{**}$, then within the range $\omega \geq \omega^{**}$ the real and imaginary parts of the LPRS $J(\omega)$ corresponding to that transfer function are monotone functions of frequency ω, with decreasing magnitudes.*

Proof. Because $|\text{Re} W_l(j\omega)|$ and $|\text{Im} W_l(j\omega)|$ are monotone decreasing functions of ω within the range $\omega \in [\omega^{**}; \infty)$, their derivatives are negative. Therefore, the functions $|\text{Re} W_l(jk\omega)|$ and $|\text{Im} W_l(jk\omega)|$, where $k = \overline{1, \infty}$, also have negative derivatives. As a result, the derivatives of the following series are negative (being sums of negative addends):

$$\frac{d \sum_{k=1}^{\infty} \frac{|\text{Im } W_l[(2k-1)\omega]|}{2k-1}}{d\omega} < 0,$$

$$\frac{d \sum_{k=1}^{\infty} |\text{Re} W_l(2k\omega)|}{d\omega} < 0,$$

$$\frac{d \sum_{k=1}^{\infty} |\text{Re} W_l[(2k-1)\omega]|}{d\omega} < 0.$$

4.4 The chattering phenomenon and its LPRS analysis 81

From the first inequality and formula (2.19), it directly follows that the absolute value of the imaginary part of the LPRS $J(\omega)$ is a monotone decreasing function of frequency ω: note that the sign of its derivative is negative. To prove the monotonicity of Re$J(\omega)$, note that the second derivatives of the functions $|\text{Re}W_l(jk\omega)|$ and $|\text{Im}W_l(jk\omega)|$, where $k = \overline{1,\infty}$, are positive (otherwise, because these functions are monotone, $W_l(j\omega)$ will not hit the origin as $\omega \to \infty$). Now we group the terms in the series (2.19) for the real part by two and find the sign of the following derivative:

$$\frac{d\left[|\text{Re}W_l[(2k-1)\omega]| - |\text{Re}W_l[2k\omega]|\right]}{d\omega}, k = \overline{1,\infty},$$

which is negative, because the first derivatives of $|\text{Re}W_l[(2k-1)\omega]|$ and $|\text{Re}W_l[2k\omega]|$ are negative and the second derivatives are positive, and the terms with higher ω have negative derivatives of smaller magnitude. Therefore, the absolute value of the real part of the LPRS $J(\omega)$ is a monotone decreasing function of frequency ω. ∎

We use the above lemmas in the following statement.

Theorem 4.4. *If the transfer function $W_l(s)$ is a quotient of two polynomials $B_m(s)$ and $A_n(s)$ of degrees m and n, respectively (4.24), then the high-frequency segment (where the above Lemma 4.2 holds) of the LPRS $J(\omega)$ corresponding with the transfer function $W_l(s)$ is located in the same quadrant of the complex plane where the high-frequency segment of the Nyquist plot of $W_l(s)$ is located, with either the real axis (if the relative degree $(n-m)$ is even) or the imaginary axis (if the relative degree $(n-m)$ is odd) being an asymptote of the LPRS.*

Proof. We prove the above theorem for an arbitrary relative degree $r = n - m \geq 1$. We note that the following infinite series have finite sums for any positive integer $r \geq 1$:

$$S_1(r) = 1 - \frac{1}{2^r} + \frac{1}{3^r} - \frac{1}{4^r} + \ldots = \left(1 - \frac{1}{2^{r-1}}\right)\varsigma(r),$$

for $r \geq 2$, and $S_1(1) = \ln 2$ for $r = 1$,

$$S_2(r) = 1 + \frac{1}{3^{r+1}} + \frac{1}{5^{r+1}} + \frac{1}{7^{r+1}} + \ldots = \left(1 - \frac{1}{2^{r+1}}\right)\varsigma(r+1),$$

where $\varsigma(r)$ is Riemann Zeta Function [37].

Also, in accordance with (2.19), the LPRS of r-th order multiple integrator is

$$J_{r-\text{int}}(\omega) = \begin{cases} (-1)^{r/2} \frac{1}{\omega^r} S_1(r) + j0 & \text{if } r \text{ is even} \\ 0 + j(-1)^{(r+1)/2} \frac{1}{\omega^r} S_2(r) & \text{if } r \text{ is odd} \end{cases} \quad (4.27)$$

with either Re$J_{r-\text{int}}(\omega)$ being zero for all odd $r = n - m$ or Im$J_{r-\text{int}}(\omega)$ being zero for even r.

We take magnitudes of the differences between the real and imaginary parts of the LPRS $J(\omega)$ and the LPRS of the multiple integrator $J_{r-\text{int}}(\omega)$ (which corresponds to the transfer function of the integrator $W_{r-\text{int}}(s) = b_m/(a_n s^r)$). Using Lemma 4.2, we derive the following inequalities:

$$|\text{Re}J(\omega) - \text{Re}J_{r-\text{int}}(\omega)| \leq \sum_{k=1}^{\infty} (-1)^{k+1} \varepsilon \left(\frac{\omega^*}{k\omega}\right)^r = \varepsilon \left(\frac{\omega^*}{\omega}\right)^r S_1(r),$$

$$|\text{Im}J(\omega) - \text{Im}J_{r-\text{int}}(\omega)| = \sum_{k=1}^{\infty} \frac{1}{2k-1} \left|\text{Im}W_l[(2k-1)\omega] - \text{Im}\frac{b_m}{ja_n(2k-1)\omega}\right|$$

$$\leq \sum_{k=1}^{\infty} \frac{\varepsilon}{2k-1} \left(\frac{\omega^*}{(2k-1)\omega}\right)^r = \varepsilon \left(\frac{\omega^*}{\omega}\right)^r S_2(r).$$

Therefore, for all $\omega \geq \omega^*$, each point of the LPRS $J(\omega)$ is located inside the rectangle, which is $2\varepsilon \left(\frac{\omega^*}{\omega}\right)^r S_1(r)$ wide by $2\varepsilon \left(\frac{\omega^*}{\omega}\right)^r S_2(r)$ high with its center at the point given by the formula for $J_{r-\text{int}}(\omega)$. The size of this rectangle is frequency-dependent, and both its dimensions tend to zero when $\omega \to \infty$. The signs of the real and the imaginary parts of the high-frequency segment of the LPRS coincide with the signs of the real and the imaginary parts of the high-frequency segment of the corresponding transfer function (Lemma 4.3). The property of either the real or the imaginary axis being an asymptote of the LPRS follows from the monotonicity and the fixed sign of the high-frequency segments of the LPRS (Lemma 4.3). ∎

The non-existence of periodic motion of finite frequency in a SM system (chattering) is therefore equivalent to the absence of points of intersection between the LPRS and the real axis (except the origin), which is addressed by the following theorem.

Theorem 4.5. *If the transfer function $W_l(s)$ is a quotient of two polynomials $B_m(s)$ and $A_n(s)$ of degrees m and n, respectively, with non-negative coefficients, then for the existence of ideal SM, it is necessary that the relative degree (n − m) of $W_l(s)$ be one or two. If the relative degree is one, then a conventional ideal SM can occur; if the relative degree is two, then the so-called asymptotic second-order SM can occur.*

Proof. The proof directly follows from Theorem 4.4. If the relative degree is higher than *two*, then the LPRS necessarily has a point of intersection with the real axis at a finite frequency. This fully coincides with classical results [1, 94]. Note 1: this does not, however, concern the case of the plant that has two or more imaginary poles (integrators). Such a system may not have a periodic solution at all. Note 2: Theorem 4.5 does not provide a sufficient condition, as a periodic motion of a finite frequency can exist even if the relative degree is *one* or *two* [19]. ∎

4.4 The chattering phenomenon and its LPRS analysis

The aforementioned theorems provide a foundation for the analysis of possible modes in a relay system. With the LPRS computed or only the transfer function available, one can easily see whether the ideal SM or chattering will occur in the SM system under analysis.

We consider a few examples of the analysis in [19] based on the relative degree of the plant transfer function along with the LPRS analysis and the modes of oscillations that may occur in a relay feedback system. In certain cases, the analysis of the relative degree is sufficient for making a conclusion about the mode in a relay system. However, a combination of the relative degree analysis and the LPRS analysis provides more reliable results.

Example 4.6. Let the plant transfer function be given by

$$W(s) = (0.5s + 1)/[(0.05s + 1)(s + 1)].$$

The relative degree of the transfer function is *one* and the LPRS fits the pattern of the first-order system. As a result, ideal SM occurs in the relay feedback system.

Example 4.7. Let the plant transfer function be given in one case as

$$W_1(s) = 1/(s^2 + s + 1)$$

and in another case be given as

$$W_2(s) = (0.005s + 1)/[(0.1s + 1)(s^2 + s + 1)].$$

Both transfer functions are of relative degree *two*. However, the LPRS corresponding to the first one does not intersect the real axis at finite frequencies (Fig. 4.4, plot #1), but the second LPRS does intersect the real axis at $\omega = 3.29 \text{s}^{-1}$ (Fig. 4.4, plot #2). A zoomed picture of the high-frequency segments shows that both LPRS have an asymptote, which is the real axis, but the second LPRS approaches the origin from the second quadrant. As a result, a second-order SM occurs if the transfer function is $W_1(s)$, and finite frequency oscillations occur if the transfer function is $W_2(s)$. Despite the fact that both linear parts are of relative degree *two*, SM cannot occur in the system with $W_2(s)$. Simulations confirm this conclusion.

Example 4.8. Let the plant transfer function be given by

$$W(s) = \frac{(0.0215s + 1)(0.00464s + 1)}{(0.1s + 1)(0.001s + 1)(s^2 + s + 1)}.$$

The transfer function is of relative degree *two*, and it includes a phase-lag-phase-lead element. As a result, the LPRS intersects the real axis from below,

84   4 Analysis of SM in the frequency domain

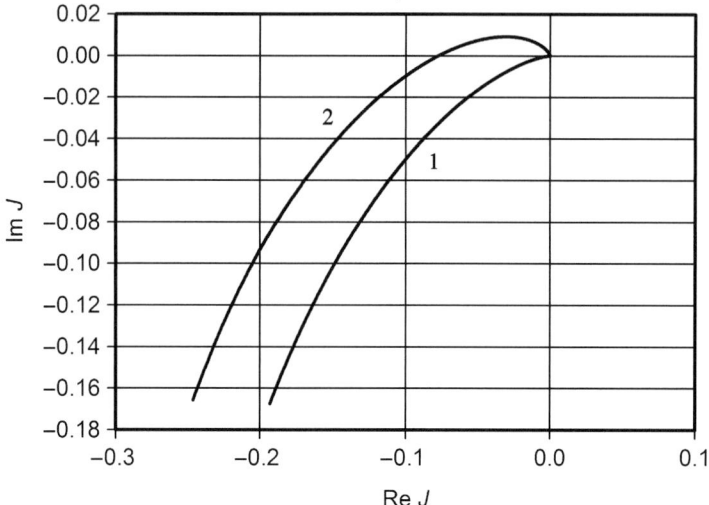

Fig. 4.4. The LPRS of Example 4.7

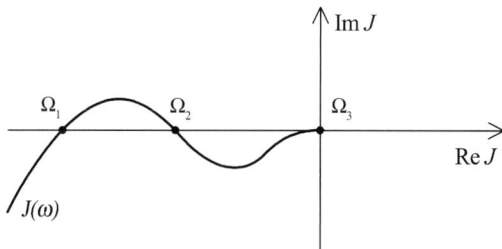

Fig. 4.5. The LPRS of Example 4.8 (qualitative behavior)

then returns to the lower half-plane and finally approaches the origin from below, with the real axis as an asymptote. The two points of the intersection are at the frequency $\Omega_1 = 3.75\text{s}^{-1}$ and at the frequency $\Omega_2 = 91.42\text{s}^{-1}$ (see the qualitative plot in Fig. 4.5). Obviously, there is one more periodic solution corresponding to the intersection at the origin: $\Omega_3 = \infty$. The frequency Ω_2 is an unstable periodic solution. However, both the other frequencies Ω_1 and Ω_3 are locally orbitally asymptotically stable solutions with certain domains of attraction. If the initial conditions are large, the process converges to the slower periodic process with frequency Ω_1. Simulations show that if the initial conditions are sufficiently small, the process converges to the periodic process of infinite frequency. Although this approach does not allow for determination of domains of attraction, it allows for prediction of the existence of two possible periodic motions.

4.5 Reduced-order and non–reduced-order models of averaged motions in a sliding mode system and input-output analysis

There are two types of input-output problems in SM control. The first one is typical of *stabilization control* and the second one applies to *servo control*. The first type of input-output problem is the determination of a response of the SM system to a constant or variable external disturbance (e.g., a static load). This problem is very similar to the analysis of the external disturbance effect on the relay feedback system. The difference between this and the relay systems analysis is that the consideration of only principal dynamics does not provide a meaningful result. The parasitic dynamics must necessarily be considered to obtain results different from the ones corresponding to ideal closed-loop performance.

The second type of input-output problem is analysis of the tracking quality of a SM servo system. Here, some specifics of the implementation of the SM servo system exist, which results in the differences between this and analysis of relay servo systems. In both cases, analysis must be done with the use of the model of averaged (over the period of the fast motions) motions in the SM system.

For the stabilization type of control, the model of the averaged motions in the system can be obtained from the original equations by replacing the nonlinear control function with the equivalent gain of the relay, in accordance with the LPRS method:

$$\dot{\mathbf{x}}_0 = \mathbf{A}\mathbf{x}_0 + \mathbf{B}u_0 + \mathbf{D}d, \tag{4.28}$$

where $\mathbf{x}_0$ and u_0 are averaged values of $\mathbf{x}$ and u, respectively, $\mathbf{x}_0 \in R^n$, $u_0 \in R^1$, $\mathbf{A}$ is an $n \times n$ matrix, $\mathbf{B}$ is an $n \times 1$ matrix, $\mathbf{D}$ is an $n \times 1$ matrix, $d \in R^1$ is the disturbance to the system, with the averaged scalar control being

$$u_0 = k_n \cdot \sigma_0, \tag{4.29}$$

where σ_0 is an averaged value of σ, $\sigma_0 \in R^1$,

$$\sigma_0 = \mathbf{S}\mathbf{x}_0, \tag{4.30}$$

and $\mathbf{S}$ is $1 \times n$ matrix. The equivalent gain k_n is computed in (2.4) with the LPRS computed as in (2.12) or (2.32). We call the system defined by equations (4.28)–(4.30) the *non–reduced-order model* of a SM system. One can see that the non–reduced-order model of the SM system is linear. For that reason, analysis of averaged motions can be done by applying the methods commonly used from linear systems theory. The objective of analysis is usually the transient process and the degree of *disturbance attenuation* by the closed-loop system.

Let us establish the relationship between the *reduced-order model* [97] and the *non–reduced-order model* (4.28)–(4.30). It is known that if the ideal SM

86   4 Analysis of SM in the frequency domain

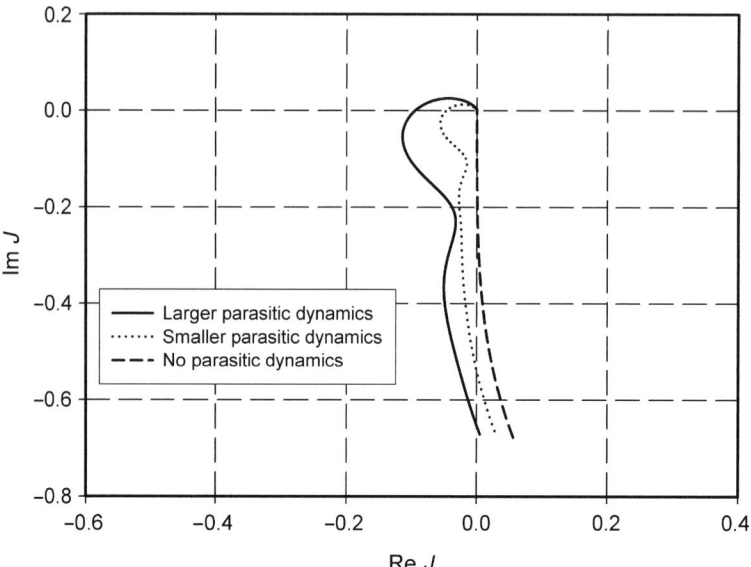

Fig. 4.6. Effect of parasitic dynamics on the LPRS

occurs, then the averaged motions in the system are described by the *reduced-order model*. As follows from the LPRS method, an ideal SM occurs if the LPRS does not have any point of intersection with the real axis except the origin (Fig. 4.6). This can only happen if the relative degree of the transfer function of the linear part is *one* or *two*. Assume that the relative degree of $W_l(s)$ is *one* and the hysteresis value b of the relay is *zero*. In that case, the straight line "$-\pi b/(4c)$" coincides with the real axis, and the LPRS (as in Theorem 4.5) approaches the origin from below, with the imaginary axis as an asymptote. Therefore, as follows from (2.3) ideal SM occurs. Yet not only does the frequency of chattering becomes infinite, but also the equivalent gain becomes *infinite*; this follows from (2.4). Figure 4.6 illustrates the transformation due to removing parasitic dynamics from the model.

We obtain equations of the SM system for this case when the equivalent gain approaches infinity. We rewrite equation (4.29) as follows:

$$u_0 = k_n \cdot \sigma_0, \quad k_n \to \infty, \tag{4.31}$$

One can see from (4.31) that the only opportunity for the averaged control u_0 to be finite occurs when the averaged sliding variable σ_0 is zero. This means that equations (4.30) and (4.31) become

$$\mathbf{S}\mathbf{x}_0 = 0. \tag{4.32}$$

Equations (4.28) and (4.32) represent the *reduced-order model*. The order of this system is $n - 1$ (recall that the order of the *non–reduced-order* model was n). Any of the state variables can be expressed through the others through (4.32) and can be excluded from equation (4.28). The *reduced-order model* describes the case of ideal sliding, which is reflected in formula (4.32). This model is *incapable* of handling the effects of external disturbance; this is illustrated by the following example of a second-order system. For the second-order system, equation (4.32) is written as

$$s_1 x_{01} + s_2 x_{02} = 0, \tag{4.33}$$

and equation (4.28) as follows (assuming that the first elements of **B** and **D** are zeros):

$$\dot{x}_{01} = x_{02}, \quad \dot{x}_{02} = a_{21} x_{01} + a_{22} x_{02} + u_0 + d. \tag{4.34}$$

From equation (4.33), x_{02} can be expressed as a function of x_{01} and substituted in equation (4.34). As a result, we obtain the following formula for the system motion

$$\dot{x}_{01} = -\frac{s_1}{s_2} x_{01}.$$

From this formula, we see that the disturbance does not affect the system motion. Therefore, we observe an ideal disturbance rejection in the *reduced-order model* of the SM control system. This, however, is not the case if the non–reduced-order model is used. Examples of this type of analysis are considered in the following section.

For the SM servo system, the variable that has to be maintained on the sliding surface is the error vector, which is defined as follows:

$$\mathbf{z} = \mathbf{f} - \mathbf{x},$$

where **f** is the input vector of the same dimension as **x**. If the plant model is given in the control canonical form, the input vector must contain the signal itself and all its derivatives up to the order $n-1$ ($f_2 = \dot{f}_1, f_3 = \dot{f}_2, f_4 = \dot{f}_3, \ldots$). We note that the necessity of higher derivatives to be available with the input signal limits the use of SM control in servo systems. The equations of the averaged motions for a SM servo system are written as follows

$$\dot{\mathbf{z}}_0 = \mathbf{A} \mathbf{z}_0 + \mathbf{B} u_0, \tag{4.35}$$

where $\mathbf{z}_0$ is the averaged value of **z** on the period of chattering, $\mathbf{z}_0 \in R^n$,

$$u_0 = k_n \cdot \sigma_0, \tag{4.36}$$

$$\sigma_0 = \mathbf{S} \mathbf{z}_0, \tag{4.37}$$

Again, the *non–reduced-order model* of the averaged motions (4.35)–(4.37) is linear, and all applicable methods of analysis can be used. However, the objective of analysis is different from the stabilization problem. The most common objective is the tracking quality of the input signal $f_1(t)$. Here we emphasize that despite the fact that a vector input is needed to organize a SM servo system, the only variable that must be tracked by the system is the first component of this vector. Tracking quality can be assessed on the basis of frequency characteristics of the closed-loop system from $f_1(t)$ to $x_1(t)$.

4.6 On fractal dynamics in sliding-mode control

There are a number of control methods that rely heavily on characteristics of the process (plant) models such as order and relative degree. Sliding mode control, high gain methods, Pontryagin's maximum principle, and approaches that use Lyapunov functions usually fall into this category. Normally, these methods use low-order (or low relative degree) plant dynamic models, which are in fact approximations. Control is usually designed for those low-order models, assuming that the effect of mismatch between the model dynamics and the actual plant dynamics is insignificant. Those models are built from first principles and usually represent only the *principal dynamics* of the process (plant). However, it is well-known from applications that besides the principal dynamics, there also exist *parasitic dynamics*. In SM control, the existence of parasitic dynamics is the cause of chattering not only in first-order SM but also in second-order SM [23] and continuous first-order and second-order SM systems [24]. The SM principle is, therefore, one of the most sensitive to parasitic dynamics control principles. Other control principles are also affected by parasitic dynamics, so that some quality deterioration due to parasitic dynamics can be expected everywhere. Therefore, this is a relatively universal problem; however, it is revealed to a higher extent in the SM system.

Let us take a closer look at the nature of principal and parasitic dynamics and their relationship. Consider a system that is supposed to control water level in a vessel. Usually the dynamics of the level in a vessel can be considered an integrator, with the valve dynamics neglected. This is very reasonable if the objective is the tank level dynamics. However, for a valve actuator designer, the emphasis is on the actuator dynamics. Most of the control valves use pneumatic actuation. At the analysis of the dynamics of the pneumatics, one neglects the electromagnet dynamics, which are not neglected by the electrical engineer who develops the electromagnet. Yet this engineer can legitimately ignore the dynamics of the electronic amplifier that is used for the control of the electromagnet. For the electronic engineer, the subject of design is the amplifier, and he can disregard the dynamics of transistors. This sequence continues to single components, junctions, and particles. According to the author's belief, this sequence is really infinite and limited only by our knowledge of nature.

The following observations can be made from the example above. (a) At each level of consideration, there always exist *principal dynamics*, which provide the main contribution to the overall dynamics, and *parasitic dynamics*, the effect of which is much smaller; these dynamics are usually neglected. (b) The connection between the principal and parasitic dynamics is serial, which is determined by the control system design principles.

We shall refer to the considered structure of process (plant) dynamics as *fractal dynamics*. This structure includes an infinite number of levels of consideration and existence of non-negligible principal dynamics and negligible parasitic dynamics on each level. The term "fractal" obviously reflects the

4.6 On fractal dynamics in sliding-mode control

similarity to fractal geometry [76]. In fractal dynamics, like in fractal geometry, one notices the property of self-similarity and the possibility of scaling at each level of consideration.

This section is devoted to the design and analysis of a model of fractal dynamics that reflects the properties of self-similarity and scaling.

Returning to the above analysis of the level control, we note that this analysis implies that the most complete description of the process dynamics is a serial connection of other dynamics, with diminishing dominant time constants. The following model in the form of a transfer function accounts for this property:

$$\Psi(T, \lambda, s) = \prod_{k=0}^{\infty} \frac{1}{\lambda^{-k} T s + 1} \tag{4.38}$$

where $\lambda > 1$ and T is a time constant. Formula (4.38) is, therefore, a serial connection of an infinite number of first-order dynamics with diminishing time constants. We call this type of dynamics first-order *fractal dynamics*. Obviously, we can also design a model of second-order *fractal dynamics*:

$$\Psi(T, \lambda, s) = \prod_{k=0}^{\infty} \frac{1}{(\lambda^{-k} T)^2 s^2 + 2\xi \lambda^{-k} T s + 1}$$

as well as other models of fractal dynamics. Below, we consider only first-order *fractal dynamics*. Consider some properties of *fractal dynamics*. The property of *self-similarity* is expressed as follows

$$\Psi(\lambda T, \lambda, s) = \frac{1}{\lambda T s + 1} \Psi(T, \lambda, s) \tag{4.39}$$

which means that an increase of the time constant by a factor of λ is equivalent to adding one more multiplier in the product (4.38). This property leads to the following function:

$$\Psi_0(\lambda, s) = \Psi(1, \lambda, s) = \prod_{k=0}^{\infty} \frac{1}{\lambda^{-k} s + 1}. \tag{4.40}$$

Then for time constants that are multiples of λ^k, the transfer function can be obtained from formula (4.39). Therefore, we rewrite formula (4.39) as follows:

$$\Psi(\lambda^n, \lambda, s) = \Psi_0(\lambda, s) \prod_{k=1}^{n} \frac{1}{\lambda^k s + 1}. \tag{4.41}$$

We prove below that despite the infinite character of the formulas for Ψ, the properties of fractal dynamics are in many ways similar to those of finite-dimensional dynamics. We analyze these properties in the frequency and time domains.

Assume that the input to the fractal dynamics is a harmonic excitation and suppose we find the harmonic response to the fractal dynamics (4.38). With

90   4 Analysis of SM in the frequency domain

a harmonic input, we replace the Laplace variable as follows: $s = j\omega$. Therefore, $\Psi(j\omega)$ is a complex function and can be represented by the magnitude component and the phase component as follows:

$$\Psi(T, \lambda, j\omega) = |\Psi(T, \lambda, j\omega)| \exp\left(j \arg\left(\Psi(T, \lambda, j\omega)\right)\right) \quad (4.42)$$

We prove that the harmonic response (4.42) provides finite attenuation of the amplitude and finite phase lag with respect to the external harmonic excitation.

Theorem 4.9. *At any finite frequency ω and $\lambda > 1$, the magnitude response of the fractal dynamics (4.38) is non-zero (finite attenuation).*

Proof. First we prove that $|\Psi_0(\lambda, j\omega)|$ provides finite attenuation to any harmonic excitation of finite frequency. We evaluate this function at $\omega = 1$.

$$\ln|\Psi_0(\lambda, j1)| \ln \prod_{k=0}^{\infty} \frac{1}{\sqrt{1+\lambda^{-2k}}} = -0.5 \sum_{k=0}^{\infty} \ln(1 + \lambda^{-2k}) \quad (4.43)$$

Using the following expansion for the logarithmic function [30]

$$\ln(1 + \alpha) = \alpha - \frac{\alpha^2}{2} + \frac{\alpha^3}{3} - \frac{\alpha^4}{4} + \ldots, \quad |\alpha| < 1 \quad (4.44)$$

we find the lower estimate of the magnitude function. Assume that $\alpha = \lambda^{-2}$. From (4.44), it follows that $\ln(1 + \alpha) < \alpha$. Thus, we write:

$$\sum_{k=0}^{\infty} \ln(1 + \lambda^{-2k}) = \sum_{k=0}^{\infty} \ln(1 + \alpha^k) < \sum_{k=0}^{\infty} \alpha^k. \quad (4.45)$$

The last formula is an infinite geometric series with sum $\frac{1}{1-\alpha}$. Therefore, given the minus sign in (4.43), we write the lower estimate for the magnitude of Ψ_0

$$\ln|\Psi_0(\lambda, j1)| > -0.5 \frac{\lambda^2}{\lambda^2 - 1}. \quad (4.46)$$

It follows from (4.46) that (4.40) converges to a finite number for any $\lambda > 1$.

Next, the function $|\Psi_0(\lambda, j\omega)|$ is obviously a decreasing function of the frequency ω, since the magnitudes of all the multipliers in (4.38) are decreasing functions. Also, it follows from (4.41) that for any frequency Ω, there exists an integer n such that the following inequality holds:

$$|\Psi(1, \lambda, j\Omega)| > |\Psi(\lambda^n, \lambda, j1)| = |\Psi_0(\lambda, j1)| \cdot \left|\prod_{k=1}^{n} \frac{1}{j\lambda^{-k} + 1}\right|$$

which is also true for an arbitrary time constant T.   ∎

Now consider the phase response of the transfer function (4.38).

4.6 On fractal dynamics in sliding-mode control

Theorem 4.10. *At any finite frequency ω and $\lambda > 1$, the phase response of the fractal dynamics (4.38) is finite (finite phase lag).*

Proof. The phase response for the transfer function (4.38) is

$$\arg \Psi(T, \lambda, j\omega) = -\sum_{k=0}^{\infty} \arctan\left(T\omega\lambda^{-k}\right). \tag{4.47}$$

From the expansion of the *arctan* function [30]

$$\arctan \alpha = \alpha - \frac{\alpha^3}{3} + \frac{\alpha^5}{5} - \frac{\alpha^7}{7} + ..., \text{ for } \alpha^2 < 1,$$

it follows that $|\arctan \alpha| < |\alpha|$ if $\alpha^2 < 1$, so we derive the following inequality:

$$\arg \Psi(T, \lambda, j\omega) = -\sum_{k=0}^{\infty} \arctan\left(T\omega\lambda^{-k}\right) > -\sum_{k=0}^{\infty}\left(T\omega\lambda^{-k}\right) = -\frac{T\omega}{1-\lambda^{-1}}. \tag{4.48}$$

In formula (4.48), the negative sign of the phase response is accounted for by using ">"; and we have used the formula of the sum for a geometric series. Formula (4.48) is only valid if $T\omega < 1$; this is not always the case. However, because $\lambda > 1$, beginning from a certain $k = k^*$, the inequality $T\omega\lambda^{-k*} < 1$ holds, so we consider two sums: one from $k = 0$ to $k^* - 1$ and one from k^* to infinity. In the second sum, we treat $T\omega\lambda^{-k^*}$ as a new value of $T\omega$. Both sums will obviously be finite.

Again, as in the case of the amplitude response, for any $\lambda > 1$ and any finite frequency, the phase response of the fractal dynamics (4.38) is finite. ∎

The phase and amplitude characteristics (Bode plot) of the fractal dynamics (4.38) for two different λ are depicted in Fig. 4.7. The Nyquist plot of (4.38) is given in Fig. 4.8 (the Nyquist plot of conventional first-order dynamics is shown for comparison, too).

One can see the linear dependence of the phase characteristic on the logarithm of frequency for higher frequencies (see Fig. 4.7). We demonstrate this.

Using formula (4.47), we find the value of $\arg \Psi(T, \lambda, j\lambda\omega)$:

$$\arg \Psi(T, \lambda, j\lambda\omega) = -\sum_{k=0}^{\infty} \arctan\left(T\lambda\omega\lambda^{-k}\right) = \arg \Psi(T, \lambda, j\omega) - \arctan(T\lambda\omega)$$

We evaluate the following limit:

$$\lim_{\omega \to \infty} \left[\arg \Psi(T, \lambda, j\lambda\omega) - \arg \Psi(T, \lambda, j\omega)\right] = -\lim_{\omega \to \infty} \arctan(T\lambda\omega) = -\frac{\pi}{2}.$$

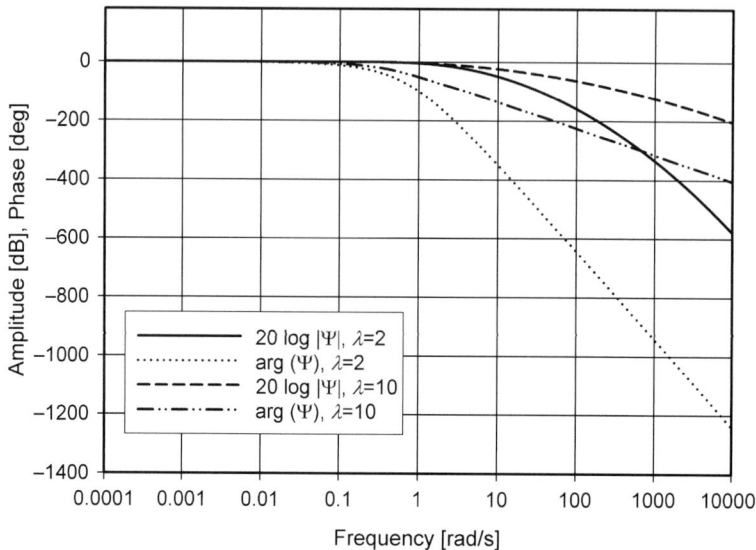

Fig. 4.7. Bode plots of fractal dynamics ($T = 1$)

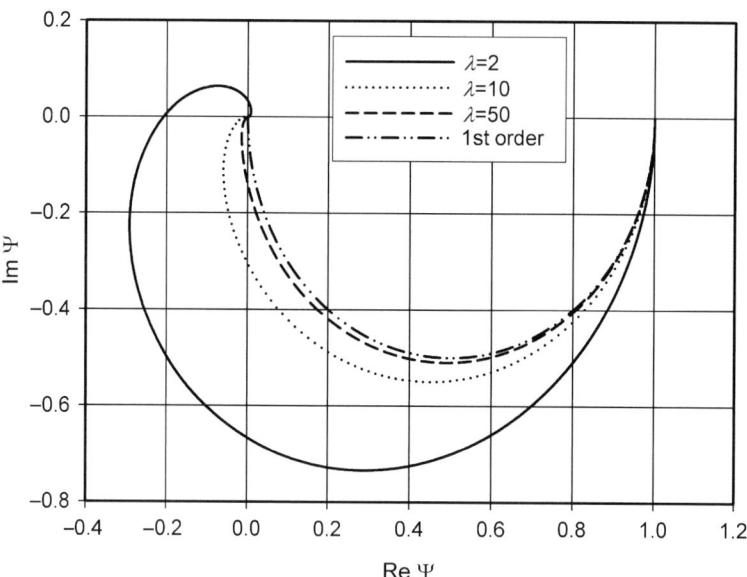

Fig. 4.8. Nyquist plots of fractal dynamics ($T = 1$)

It follows from the last formula that the slope of the phase characteristic of the fractal dynamics at high enough frequencies is $90°/\log\lambda$ per decade. Therefore, for $\lambda = 2$, it is $299°$/decade and for $\lambda = 10$ it is $90°$/decade (Fig. 4.7).

4.6 On fractal dynamics in sliding-mode control

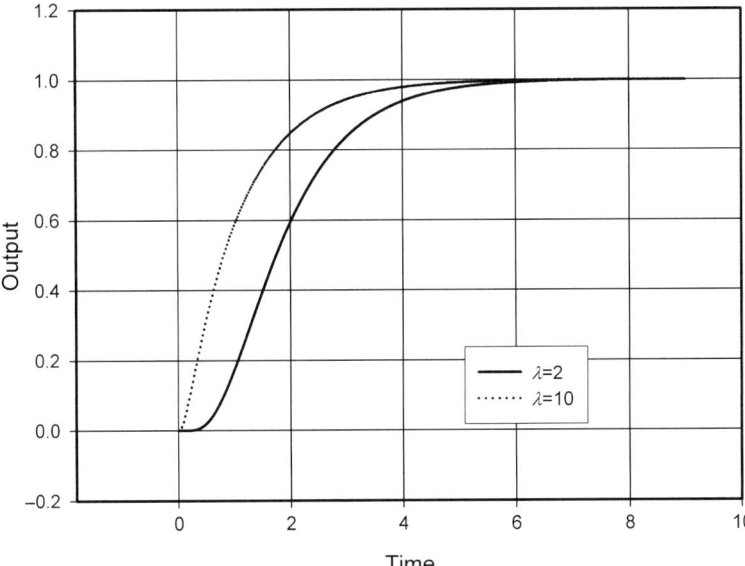

Fig. 4.9. Step response of fractal dynamics ($T = 1$)

Regarding the time-domain characteristics of the fractal dynamics, the most important is the step response of the dynamics (4.38). It is difficult to obtain analytical formulas for the step response. For that reason, we approximate this function numerically via multiple integration (as per (4.38)) of the time evolution of the response of first-order dynamics $W(s) = 1/(Ts + 1)$: $h(t) = 1 - \exp(-t/T)$. The step response of (4.38) for two different values of λ is given in Fig. 4.9. In Fig. 4.9, the step input is applied at $t = 0$. One can see that the effect of the fractal dynamics is similar to that of time delay (see the plot for $\lambda = 2$), as there is an initial time interval when the output stays almost equal to zero. However, fractal dynamics are minimal-phase, which makes them different from time delay. The length of this initial interval depends on the parameter λ. One can see that the length of the initial time interval for $\lambda = 2$ is larger than for $\lambda = 10$. The existence of such an initial time interval in a step response of real plants was first identified in [109] many years ago. In fact, the response in Fig. 4.9 is often approximated by a first-order plus dead-time model for the purpose of process identification and controller tuning.

We compute the LPRS for the dynamics (4.38) using formula (2.19). The LPRS plots for two different values of λ are given in Fig. 4.10. We have already shown that the point of the LPRS corresponding to zero frequency is always $(0.5; -j\pi/4)$ for all non-integrating plants with unity static gains. One can see from Fig. 4.10 that it is also the case for the fractal dynamics. Also, similarly to the Nyquist plots, the LPRS plots have a spiral shape around the origin with the frequency tending to infinity.

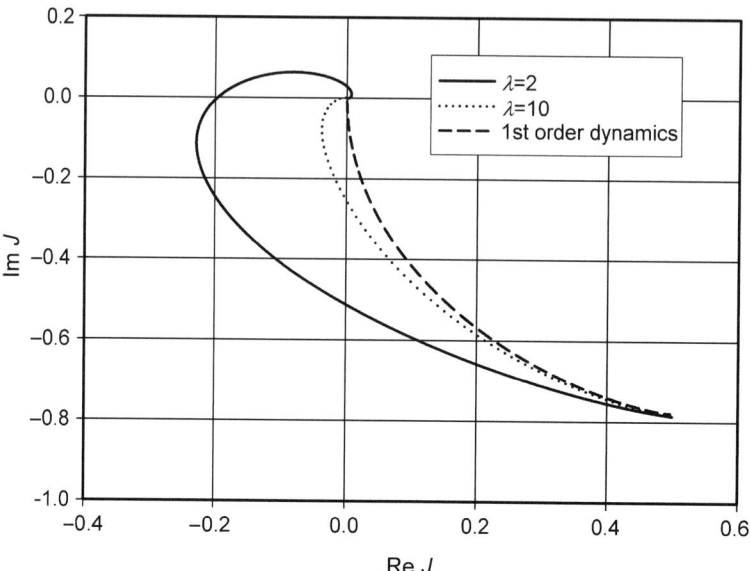

Fig. 4.10. The LPRS of fractal dynamics ($T = 1$)

It is well-known and shown above that chattering in SM control systems is caused by the inevitable existence of *parasitic dynamics*, which exist along with the *principal dynamics* of the plant. The *principal dynamics* are the dynamics of the model of the plant and of the sliding surface. Yet, for implementation of the designed control algorithms, devices such as actuators and sensors are needed, which brings into the system certain *parasitic dynamics* not accounted for in the SM control design. Due to the SM design principle that involves the use of the sliding surface, the relative degree of the *principal dynamics* is always *one*. If no *parasitic dynamics* are present, the LPRS has the shape similar to the one depicted in Fig. 4.10 for the first-order dynamics (dashed line). In this case, the LPRS would not have any points of intersection with the real axis except the origin. However, in any actual application, *parasitic dynamics* always exist along with the principal ones. Let us assume that the parasitic dynamics are fractal (4.38). In that case, the *principal dynamics* are connected in series with the *parasitic dynamics*, which are fractal. The LPRS of the overall dynamics then has a shape similar to that depicted in Fig. 4.10 for $\lambda = 2$ or $\lambda = 10$ (spiral motion with increased omega). In particular, a point of intersection between the LPRS and the real axis always exists. Moreover, this point is not the origin, which means that finite-frequency oscillations (chattering) will necessarily occur in a real SM system.

Furthermore, due to the spiral shape of the LPRS, the point of first intersection is always located on the left half-plane, and the following equality holds: $\mathrm{Re}J(\Omega) < 0$. As a result, the equivalent gain of the relay always a finite positive value, $0 < k_n < \infty$, which results in a deteriorated closed-loop

4.7 Examples of chattering and disturbance attenuation analysis

Consider three examples that illustrate the analysis of chattering and disturbance attenuation analysis.

Example 4.11. Consider the equations of a spring-loaded cart with viscous output damping on an inclined plane, which can be written as follows,

$$\dot{x} = x_2, \dot{x}_2 = -x_1 - x_2 + u_a + d,$$

where x_1 is the linear displacement of the cart, x_2 is the linear velocity, u_a is the force developed by the actuator, and d is the disturbance (projection of the gravity onto the inclined plane). The goal is to stabilize the cart at the point corresponding to zero displacement. We design the switching surface (line) as $x_1 + x_2 = 0$ and the control as a relay control that can make the point $x = 0$ an asymptotically stable equilibrium point of the closed-loop system under the applied disturbance $d = -1 : u = -4\text{sign}(x_1 + x_2)$. Suppose that the force u_a is developed by an actuator with the second-order dynamics

$$T_a^2 \ddot{u}_a + 2\xi_a T_a \dot{u}_a + u_a = u,$$

where $T_a = 0.01\text{s}^{-1}$, $\xi_a = 0.5$. Clearly, the system should exhibit oscillations due to the actuator presence. Finding the frequency and the amplitude of those oscillations is one of our goals. Another goal is an assessment of the disturbance effect. In the case of ideal sliding, even if the disturbance is applied, the trajectory tends to the origin. In the case of non-ideal sliding (due to the actuator presence), the trajectory does not tend to the origin. We write an expression for the transfer function of the linear part $W_l(s) = (s+1) \cdot W_a(s) \cdot W_p(s)$, where $W_a(s) = 1/(T_a^2 s^2 + 2\xi_a T_a s + 1)$, $W_p(s) = 1/(s^2 + s + 1)$. We compute the LPRS for $W_l(s)$ as per (2.19), and plot it on the complex plane (Fig. 4.11).

We find the point of intersection between the LPRS and the real axis. This point corresponds to the frequency $\Omega = 99.27\ s^{-1}$, which is the frequency of chattering in the system. The real part of the LPRS at this point is $\text{Re}J(\Omega) = -0.00946$, and the equivalent gain of the relay (according to formula (2.4)) is $k_n = 52.8$. As a result, the model of the averaged (slow) motions is written as follows (subscript '0' denotes the slow component of respective variables):

$$\begin{cases} \dot{x}_{01} = x_{02}, \\ \dot{x}_{02} = -x_{01} - x_{02} + u_{0a} + d, \\ \ddot{u}_{0a} = (u_0 - 2\xi_a T_a \dot{u}_{0a} - u_{0a})/T_a^2, \\ u_0 = -k_n \sigma_0, \\ \sigma_0 = x_{01} + x_{02}. \end{cases} \quad (4.49)$$

96   4 Analysis of SM in the frequency domain

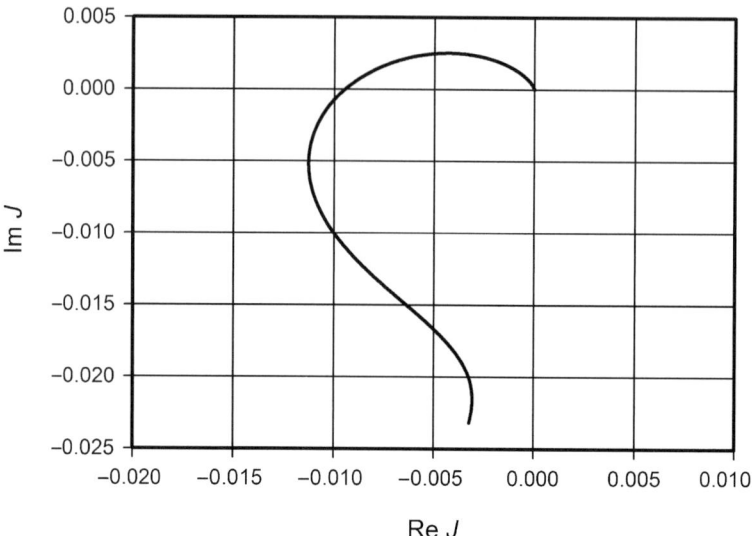

Fig. 4.11. The LPRS of the linear part (actuator, plant, and sliding surface) of the system in Example 4.9

The reduced order model can be obtained from (4.49) as a limiting case: if the equivalent gain is set to infinity: $k_n \to \infty$ (this results in $\sigma_0 = 0$ and, consequently, in $x_{01} = -x_{02}$, the condition of ideal sliding). However, the actual value of the equivalent gain computed above is finite. For that reason the *non–reduced-order model* provides higher accuracy in comparison to the *reduced-order model*. Because the transient processes in both the *reduced-order model* and the *non–reduced-order model* look alike, the advantage of the *non–reduced-order model* can be best demonstrated if an external disturbance is applied to the system, and the effect of this disturbance is of interest.

In this example, the equivalent gain k_n is constant, which follows from the time-invariance of the linear part (this results in the same point of intersection between the LPRS with the real axis). For that reason, the effect of the applied disturbance is identical in the transient and the steady-state modes, and the analysis of disturbance attenuation can be carried out with the use of techniques for linear systems. We analyze the disturbance attenuation. In a steady state, there exists periodic motion of frequency Ω with the center $(x_{01}, 0)$ where $x_{01} = d/(1 + k_n) = -0.018$, which can be considered a disturbance rejection measure. That means that in a steady state, the cart exhibits oscillations around the point $x_{01} = -0.018$, with the frequency $\Omega = 99.27\ s^{-1}$ and the amplitude of the fundamental frequency component: $A_{x1} = 4c/\pi |W_a(j\Omega) \cdot W_p(j\Omega)| = 5.19 \cdot 10^{-4}$. The simulations of the original equations provide the following results (respective variables have the subscript *sim*). The frequency of chattering is $\Omega_{sim} = 99.21 s^{-1}$, and the output

4.7 Examples of chattering and disturbance attenuation analysis

averaged steady-state value is $x_{01sim} = -0.017$, which closely match the frequency domain analysis.

Example 4.12. Consider an example similar to the previous one but with an actuator given as fractal dynamics.

Consider the same equations of the spring-loaded cart with viscous output damping on the inclined plane as in the previous example, and the same control law. The applied disturbance is again $d = -1$.

Now suppose that the force u_a is developed by an actuator the dynamics of which are fractal with dominant time constant $T_a = 0.01\text{s}^{-1}$ and parameter $\lambda = 5$. Therefore, the transfer function of the actuator is

$$W_a(s) = \Psi(0.01, 5, s) = \prod_{k=0}^{\infty} \frac{1}{5^{-k}0.01s + 1}.$$

We find the frequency of chattering and the displacement of the cart due to the disturbance.

We write an expression for the transfer function of the linear part

$$W(s) = (s+1)W_a(s) \cdot W_p(s),$$

where $W_p(s) = 1/(s^2 + s + 1)$.

We compute the LPRS for $W(s)$ and plot it on the complex plane (Fig. 4.12). We find the point of intersection between the LPRS and the real

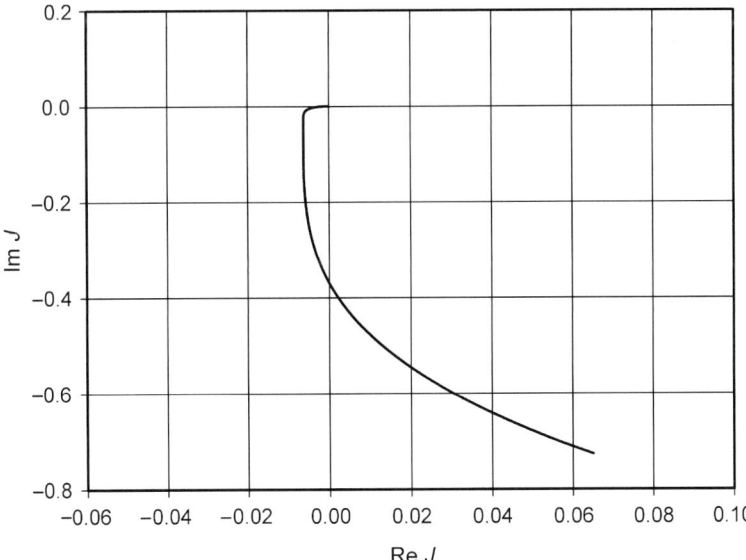

Fig. 4.12. The LPRS of the linear part (actuator, plant, and sliding surface)

axis. This point corresponds to the frequency $\Omega = 189.4\text{s}^{-1}$, which is the frequency of chattering in the system. The real part of the LPRS at this point is $\text{Re}\, J(\Omega) = -0.00194$ and the equivalent gain of the relay (according to formula (2.4)) is $k_n = 257.7$. As a result, the non–reduced-order model of the slow motions can be written as follows (subscript '0' denotes the slow component of respective variables),

$$\begin{cases} \dot{x}_{01} = x_{02} \\ \dot{x}_{02} = -x_{01} - x_{02} + u_{0a} + d \\ u_0 = -k_n \sigma_0 \\ \sigma_0 = x_{01} + x_{02} \\ u_{0a}(s) = W_a(s) \cdot u_0(s) \end{cases}$$

where the last formula is written in the Laplace domain (due to the infinite-dimensional character of the dynamics).

The reduced-order model can be obtained as a limiting case: if the time constant $T_a \to 0$ and consequently the equivalent gain $k_n \to \infty$, then this results in $\sigma_0 = 0$ and, consequently, in $x_{01} = -x_{02}$: i.e., the condition of ideal sliding. We note that the reduced-order dynamics are the same as in the previous example.

We analyze the disturbance attenuation. In a steady state, there exists oscillations of frequency Ω with the center $(x_{01}, 0)$ where $x_{01} = d/(1 + k_n) = -0.0039$, which can be considered a disturbance rejection measure. Therefore, in a steady mode, the cart exhibits oscillations around the point $(-0.0039, 0)$, with the frequency $\Omega = 189.4\text{s}^{-1}$.

A comparison with the results of the previous example shows high sensitivity of chattering parameters and closed-loop performance with respect to the parasitic (actuator) dynamics. It is apparent that the faster the actuator is, the higher the frequency and the smaller the amplitude of chattering, and the better the closed-loop performance of the SM system is.

Example 4.13. In this example, we carry out the transformation of the original SM system to the equivalent relay feedback form. Consider the pendulum equation (we assume the angles are small and, consequently, the sine function is equal to the argument):

$$\dot{x}_1 = x_2, \tag{4.50}$$

$$\dot{x}_2 = -x_1 - x_2 + u_a - \delta_1 \tag{4.51}$$

where $x_1 = \theta - \delta_1$, $x_2 = \dot{\theta}$, u_a is the torque input, and θ is the pendulum angle. The goal is to stabilize the pendulum at the angle $\theta = \delta_1$. In [62], a state feedback control is designed as a sum of an equivalent control and a switching control, which makes $\mathbf{x} = 0$ an asymptotically stable equilibrium point of the closed-loop system

$$u = x_1 + \delta_2 - 4(1 + |x_2|)\text{sign}(x_1 + x_2) \tag{4.52}$$

where δ_2 is an estimate of δ_1, which is used in the equivalent control component of the designed control. Suppose that the torque u_a is developed by a dynamic actuator, and $u_a \neq u$:

$$T_a^2 \ddot{u}_a + 2\xi_a T_a \dot{u}_a + u_a = u, \ T_a = 0.01\text{s}, \ \xi_a = 0.5.$$

We transform system (4.50)–(4.52) into an equivalent relay system with variable relay output (Fig. 4.4). We take the equivalent control term x_1 of (4.52) into account within the transfer function of the linear part by closing the respective feedback. As a result, the transfer function of the linear part can be written as follows (the multiplier "4" is transposed to the linear part that makes the relay amplitude equal to $(1 + |x_2|)$),

$$W_l(s) = 4(s+1)[W_a(s) \cdot W_p(s)/(1 - W_a(s) \cdot W_p(s))],$$

where $W_a(s) = 1/(T_a^2 s^2 + 2\xi_a T_a s + 1)$, $W_p(s) = 1/(s^2 + s + 1)$.

We bring the control function to the form of formula (4.7) and neglect the continuous term:

$$u = \delta_2 - (1 + x_{01})\text{sign}\,\sigma.$$

Obviously, stabilization of the pendulum at a non-zero angle involves a disturbance applied to the pendulum. In the case of ideal sliding, even if $\delta_2 \neq \delta_1$, the trajectory tends to the origin. In the case of non-ideal sliding, due to the actuator presence, if $\delta_2 \neq \delta_1$, there is an uncompensated disturbance $D = -\delta = \delta_2 - \delta_1$ and the trajectory does not tend to the origin. In Fig. 4.13, neither the disturbance δ_1 nor the control component δ_2 are shown; the uncompensated part of the disturbance is shown instead.

We calculate and plot the LPRS for $W_l(s)$ (Fig. 4.14). We find the point of intersection between the LPRS and the real axis. This point corresponds

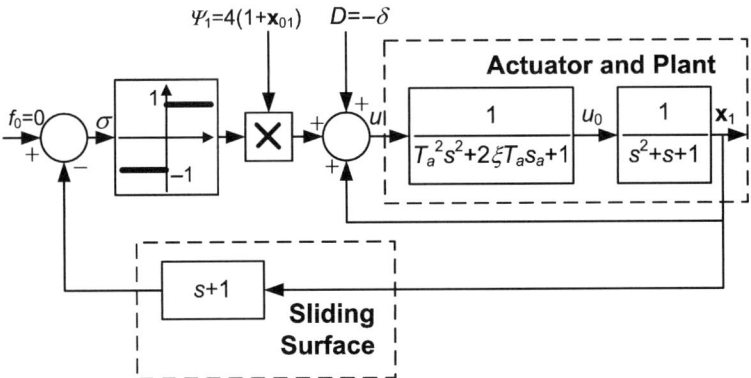

Fig. 4.13. Equivalent relay system

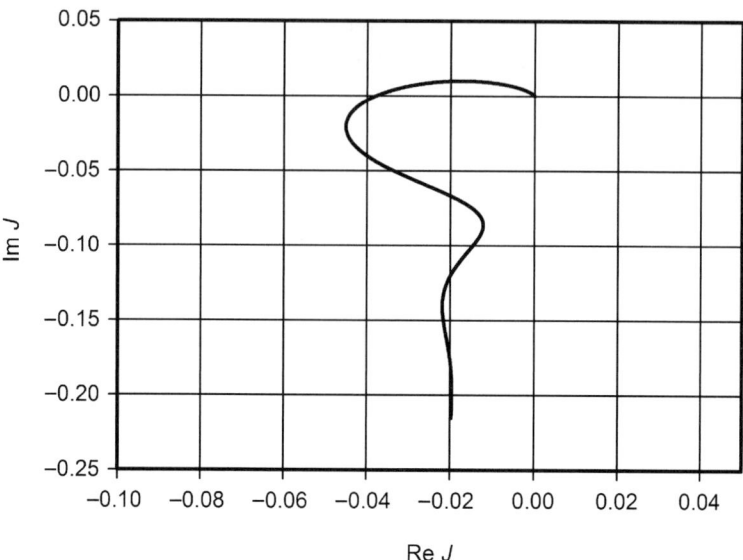

Fig. 4.14. LPRS for transfer function $W_l(s)$ in Example 4.10

to the frequency $\Omega = 99.3s^{-1}$, which is the frequency of chattering in the system.

The real part of the LPRS at this point is $\text{Re}J(\Omega) = -0.0354$, and the equivalent gain of the relay (according to formula (2.4)) is $k_n = 14.1$. As a result, the model of the slow motions can be written as follows (subscript "0" denotes the slow component of respective variables):

$$\begin{cases} \dot{x}_{01} = x_{02} \\ \dot{x}_{01} = -x_{01} - x_{02} + u_{0a} \\ \ddot{u}_{0a} = (u_0 - 2\xi_a T_a \dot{u}_{0a} - u_{0a})/T_a^2 \\ u_0 = x_{01} - \delta - k_n \sigma_0 \\ \sigma_0 = x_{01} + x_{02}. \end{cases} \quad (4.53)$$

In (4.53), the equivalent gain k_n does not vary; its value is calculated above as $k_n = 14.1$. The reduced-order model can be obtained from (4.53) as a limiting case if the equivalent gain is set to infinity: $k_n \to \infty$, which results in $\sigma_0 = 0$ and, consequently, in $x_{01} = -x_{02}$, the condition of ideal sliding. We note that the actual value of the equivalent gain is finite and, moreover, it is not a large number. For that reason, in this case, the non–reduced-order model provides additional accuracy compared to the reduced-order model.

Because the equivalent gain k_n does not depend on the amplitude of the relay, the effect of the applied disturbance is identical in the transient and the steady-state modes, and analysis of disturbance attenuation can be carried out with linear systems techniques. Let us take $\delta = 0.1$ and analyze the

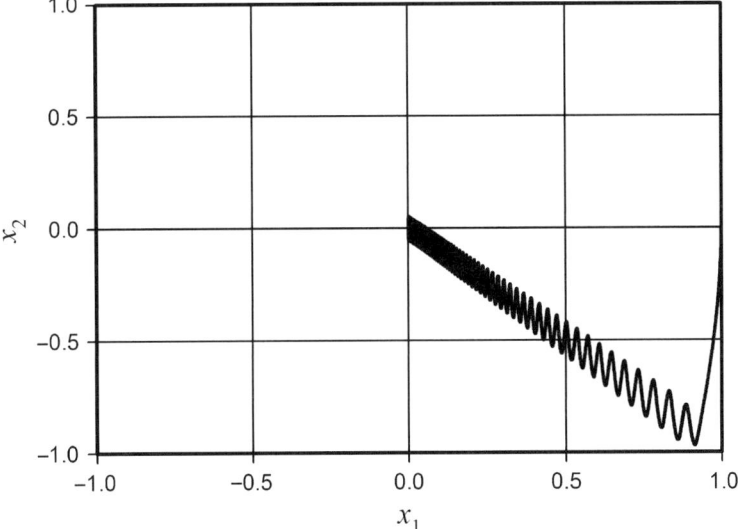

Fig. 4.15. State trajectory from initial point (1,0)

disturbance attenuation. In the steady-state, there exist oscillations of frequency Ω centered at $(x_{01}, 0)$ where $x_{01} = \delta/(4\,k_n) = 0.0018$, which is a disturbance rejection measure. This means that in a steady state, the pendulum exhibits oscillations around the point $(0.1 - 0.0018, 0) = (0.0982, 0)$, with frequency $\Omega = 99.3\text{s}^{-1}$ and amplitude of the fundamental frequency component

$$A_{x1} = \frac{4c}{\pi}|W_{a-p}(j\Omega)| = 5.20 \cdot 10^{-4} \text{ rad},$$

where $W_{a-p}(s) = 4[W_a(s) \cdot W_p(s)/(1 - W_a(s) \cdot W_p(s))]$, $c = 1 + \lim_{t \to \infty} x_{01} \approx 1$.

We run a simulation of the original equations and compare the results with the frequency domain analysis. The transient process in the state space is presented in Fig. 4.15. The frequency of chattering determined from the simulations is $\Omega_{sim} = 99.7\text{s}^{-1}$, and the output average steady-state value is $x_{01sim} = 0.0019$ rad, which closely matches the frequency-domain analysis.

4.8 Conclusions

In this chapter, we present a frequency-domain approach to analysis of self-excited and forced motions in a sliding mode system with parasitic dynamics. We demonstrate that even if the control algorithm is not a relay one, both the self-excited and the forced motions can be analyzed as motions in a certain "equivalent" relay system. We give a methodology of bringing an original sliding mode system to the equivalent relay form. We show that chattering is a self-excited motion, which can be analyzed as a limit cycle in the equivalent relay system, and the forced motions can be derived from the solution of a certain equivalent *linear* system with a time-varying or constant equivalent gain of the relay. The order of the system for the forced motions analysis is equal to the order of the original system. The model of averaged motions in the SM system that retains the order of the original system is referred to as the *non–reduced-order model*. We show how the LPRS method can be used for analysis of the equivalent relay system.

We prove that in the limiting case corresponding with the absence of parasitic dynamics, both the frequency of chattering and the equivalent gain of the relay approach infinity. This in turn leads to the traditional *reduced-order model* of a SM control system. The examples of analysis of chattering and disturbance attenuation illustrate the methodology of the LPRS analysis of SM control systems.

5

Performance analysis of second-order SM control algorithms

5.1 Introduction

Preliminaries. The SM control approach was developed in the late 1950s, and the very first implementations of the sliding mode control technique showed that the real sliding mode exhibited chattering, which is the most problematic issue in sliding mode control applications [99, 105].

Three main approaches to chattering elimination and attenuation in SM control systems were proposed in the 1980s:

- *The use of saturation control* instead of discontinuous control [31, 90]. This approach allows for control continuity but cannot restrict the system dynamics onto the switching surface. It only ensures the convergence to a boundary layer of the sliding manifold, the size of which is defined by the slope of the saturation function.
- *The observer-based approach* [27]. This method allows for bypassing the plant dynamics by the chattering loop. The approach reduces the problem of robust control to the problem of exact robust estimation and consequently can lead to the deterioration of robustness with respect to plant uncertainties, due to the mismatch between the observer and plant dynamics [105].
- *The high-order sliding modes* [40, 66]. These modes allow for finite-time convergence to zero of not only the sliding variable but its derivatives, as well. This approach was actively developed over the past two decades ([11–14, 42, 68, 80, 86]) as not only a means of chattering attenuation but also a means of robust control of plants of relative degree *two* and higher. Theoretically, the *r-th* order sliding mode totally suppresses chattering in the system with a plant of relative degree r (this, however, does not mean that chattering in a real system can be suppressed; recall the intrinsic existence of parasitic dynamics). Yet, any model is an approximation of a real system and cannot fully account for parasitic dynamics; consequently, the chattering effect cannot be avoided [23, 24].

In previous chapters we showed that in the conventional SM control system, parasitic dynamics cause chattering and deteriorated closed-loop performance. Let us analyze the system controlled by a second-order SM control algorithm and prove that the same effects exist in those systems, too, due also to the presence of parasitic dynamics.

Among higher-order sliding modes, only the second-order sliding modes (SOSM) feature prominently in applications. An important application of SOSM is in the SM observers instead of the conventional first-order SM controllers. SOSM controllers do not eliminate chattering but usually provide smaller amplitudes of chattering, which makes the subsequent low-pass filtering easy and may enhance the overall performance of the observer [23–25].

SOSM are realized not as simple on-off algorithms that switch the control depending on the sign of the sliding variable but as more complex algorithms that, compared with the conventional SM, offer *advance switching* of the control ("twisting algorithm" [66], "sub-optimal algorithm" [15], "prescribed control law" [68]). Sometimes, if the principal dynamics is of relative degree *one*, in addition to the *advance switching*, SOSM offer *control smoothing* by including an integrator in series with the discontinuous nonlinearity ("super-twisting algorithm" [68], "twisting-as-a-filter algorithm" [66]).

In this chapter, we analyze the sub-optimal algorithm using first the DF and next the LPRS methods.

5.2 Sub-optimal algorithm

The sub-optimal algorithm was proposed in [15], and further studied in [11, 12] and analyzed with respect to the control of mechanical systems in [13]. Consider an application of this algorithm to the control of a linear plant. As before, assume that the plant is given as follows,

$$\dot{\mathbf{x}} = \mathbf{A}\mathbf{x} + \mathbf{B}u$$
$$y = \mathbf{C}\mathbf{x} \qquad (5.1)$$

where $\mathbf{A} \in R^{n \times n}$, $\mathbf{B} \in R^{n \times 1}$, and $\mathbf{C} \in R^{1 \times n}$ are matrices, $\mathbf{x} \in R^n$, $y \in R^1$. The "generalized sub-optimal" SOSM control algorithm is given as follows [15],

$$u = -c \operatorname{sign}(y - \beta y_{Mi}) \qquad (5.2)$$

where c and β are controller parameters (constants) and y_{Mi} is the latest "singular point" of y, i.e., the value of y at the most recent time instant t_{Mi} ($i = 1, 2, \ldots$) such that $\dot{y}(t_{Mi}) = 0$. We assume system (5.1) describes the combined *principal* and *parasitic dynamics*.

Let us assume that the steady-state behavior of the system (5.1), (5.2) is a periodic, unimodal symmetric limit cycle with zero mean and show that the motion under assumption can exist. The sequence of singular points of

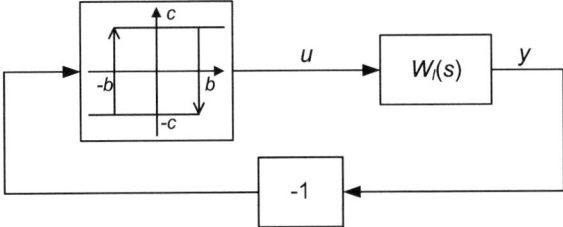

Fig. 5.1. Relay servo system representation of the sub-optimal algorithm

the variable $y(t)$ is then an alternating sequence of positive and negative values of the same magnitude: $y_M^p, -y_M^p$ (where "p" stands for periodic). The switching of the control occurs at the instants when the plant output $y(t)$ is equal either to βy_M^p or to $-\beta y_M^p$. This corresponds to the hysteretic relay characteristic of the controller (Fig. 5.1).

5.3 Describing function analysis of chattering

With this representation, the DF method [8, 50] can be conveniently used to analyze a system with the sub-optimal algorithm. The usual assumption for applicability of the DF method is that the linear part (the combined *principal* and *parasitic dynamics*) satisfies the filtering hypothesis. The DF analysis is a simple approach that can provide, in most cases, a sufficiently accurate estimate of the frequency and the amplitude of possible periodic motion. The main difference between this case and the conventional relay system (even having a relay with a negative hysteresis value) is that the hysteresis value βy_M^p is actually *unknown*. To solve this problem, we assume during a periodic motion, the extreme values of the output coincide, in magnitude, with its amplitude. Therefore, y_M^p is actually the unknown amplitude of the periodic motion. The DF of the relay with a negative hysteresis is given as in [8]:

$$N(a_y) = \frac{4c}{\pi a_y}\sqrt{1 - \frac{b^2}{a_y^2}} + j\frac{4bc}{\pi a_y^2} \tag{5.3}$$

where $b = \beta y_M^p$ is a half of the hysteresis, c is the relay amplitude, and $a_y = y_M^p$ is the amplitude of the harmonic input to the relay. We exploit the relationships between the hysteresis value and the oscillation parameters to obtain the following expression for the DF of the generalized sub-optimal SOSM algorithm:

$$N(a_y) = \frac{4c}{\pi a_y}\left(\sqrt{1 - \beta^2} + j\beta\right). \tag{5.4}$$

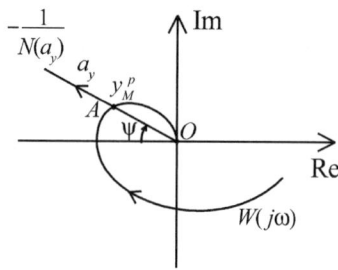

Fig. 5.2. DF analysis of self-excited oscillations

A periodic solution can be found from the harmonic balance equation $N(a_y)W(j\omega) = -1$ [8], which can be rewritten as

$$W(j\omega) = -\frac{1}{N(a_y)} \qquad (5.5)$$

where the negative reciprocal of the DF (5.4) is as follows:

$$-\frac{1}{N} = -\frac{\pi a_y}{4c}\left(\sqrt{1-\beta^2} - j\beta\right). \qquad (5.6)$$

As usual, the periodic solutions correspond to the points intersection between the $-1/N(a_y)$ and $W(j\omega)$ loci in the complex plane. The locus (5.6) is a straight line that begins at the origin and makes an angle of $\arcsin(\beta)$ with the horizontal axis, as depicted in Fig. 5.2.

Therefore, periodic motion may occur if, at some frequency $\omega = \Omega$, the phase characteristic of the actuator-plant frequency response $W(j\omega)$ is equal to $-180^0 - \arcsin(\beta)$. If this requirement is fulfilled, so that intersection between the two plots occurs, then the frequency and the amplitude of the periodic solution can be derived from the "cross-over" frequency Ω and from the magnitude $|OA|$ in Fig. 5.2, respectively. An intersection point will certainly exist if the overall relative degree of the combined actuator-plant degree (combined *principal* and *parasitic dynamics*) is three or higher. From the DF analysis of the generalized sub-optimal algorithm, we can also conclude that the frequency of self-excited oscillations (chattering) in the system with this SOSM is always higher than the frequency of the oscillations in the system with the conventional relay controller (ideal relay). This frequency increase depends on the value of the parameter β.

5.4 Exact frequency-domain analysis of chattering

The DF analysis given above provides a simple and systematic, but approximate, evaluation of magnitude and frequency of periodic motions in the closed-loop system (5.1)–(5.2) driven by the generalized sub-optimal SOSM algorithm. As noted above, despite the fact that in a periodic motion the generalized sub-optimal algorithm acts as the relay system, the hysteresis value of this relay depends on the amplitude of the oscillation and is, therefore, unknown. Hence a conventional application of the LPRS method is impossible. Nevertheless, an exact solution can be obtained by applying the following frequency-domain approach.

We introduce the following complex function $\Phi(\omega)$:

$$\Phi(\omega) = -\sqrt{[a_y(\omega)]^2 - y^2\left(\frac{\pi}{\omega},\omega\right)} + jy\left(\frac{\pi}{\omega},\omega\right) \qquad (5.7)$$

where $y\left(\frac{\pi}{\omega},\omega\right)$ is the value of the system output at the time instant when the relay switches from $-c$ to c (π/ω is half a period for the periodic motion and $t = 0$ is assumed, without loss of generality, to be the time of the relay switch from c to $-c$), and $a_y(\omega)$ is the amplitude of the plant output in the assumed periodic motion of frequency ω:

$$a_y = \max_{t \in [0,T]} |y(t,\omega)|. \qquad (5.8)$$

$y(t,\omega)$ can be computed by means of its Fourier series

$$\begin{aligned} y(t,\omega) &= \frac{4c}{\pi}\sum_{k=1}^{\infty}\frac{1}{k}\sin(\tfrac{1}{2}\pi k)\sin[k\omega t + \varphi(k\omega)]L(\omega k) \\ &= \frac{4c}{\pi}\sum_{k=1}^{\infty}\frac{(-1)^{k+1}}{2k-1}\sin[(2k-1)\omega t + \varphi((2k-1)\omega)]\cdot L((2k-1)\omega) \end{aligned} \qquad (5.9)$$

where $\varphi(k\omega) = \arg W(jk\omega)$, $L(k\omega) = |W(jk\omega)|$ are the phase and magnitude of $W(j\omega)$ at the frequency $k\omega$, respectively.

The frequency-dependent variable $a_y(\omega)$ can be computed using (5.8) and (5.9) and $y(\frac{\pi}{\omega},\omega)$ as the imaginary part of the LPRS (with coefficient $\frac{4c}{\pi}$) or through the Fourier series (5.9). As a result, the function $\Phi(\omega)$ has the same imaginary part as the Tsypkin locus [94] (or as the imaginary part of the LPRS with a coefficient), and the magnitude of the function $\Phi(\omega)$ at the intersection point represents the amplitude of the periodic solution.

Having computed the function $\Phi(\omega)$, we carry out analysis of possible periodic motions in the same way as it was done above via the DF technique, simply replacing the Nyquist plot of $W(j\omega)$ with the function $\Phi(\omega)$ given by (5.7).

The methodology of the exact frequency-domain analysis is the same as that of the DF analysis. Again, the point of intersection of the straight line drawn through the origin and at an angle with the horizontal axis equal to $\arcsin\beta$, as depicted in Fig. 5.2, and of the locus $\Phi(\omega)$ gives the frequency and the amplitude of the periodic motion (chattering). The qualitative conclusions regarding chattering are the same as in the DF analysis: *self-excited oscillations (chattering) always exist if the relative degree of the linear part transfer function is higher than two*, and *the frequency of chattering in the system with the generalized sub-optimal algorithm is always higher than the frequency of chattering in the system with the conventional relay controller (ideal relay).* The example below illustrates the methodology of analysis and these conclusions.

5.5 Describing function analysis of external signal propagation

The autonomous properties of the generalized sub-optimal algorithm were investigated above. We showed that it is superior in some respects to conventional relay control. However, pure autonomous modes never occur in real systems due to the existence of external disturbances and servo modes. Therefore, the analysis of transfer properties of systems controlled by SOSM controllers and, in particular, by the generalized sub-optimal algorithm is important.

Let us apply an approach similar to the one used for the analysis of transfer properties of the relay systems. First, we apply a constant input to the closed-loop system controlled by the generalized sub-optimal algorithm; after that, considering this constant input an infinitesimally small value, we determine the equivalent gain of the generalized sub-optimal algorithm; and finally extend the obtained results to the case of variable inputs (slow and relatively fast).

Suppose the system is controlled by the generalized sub-optimal algorithm as in the block diagram (Fig. 5.3), where the generalized sub-optimal algorithm is applied to the error signal.

An external constant input f_0 is applied to the system. Let us assume that the switching happens according to the following equation:

$$u = c \, \text{sign} \, (\sigma - \beta \sigma_{Mi}), \qquad (5.10)$$

where σ_{Mi} is the latest "singular point" of σ, i.e., the value of σ at the most recent time instant t_{Mi} $(i = 1, 2, \ldots)$ such that $\dot{\sigma}(t_{Mi}) = 0$, so that the switching instants depend on the amplitude of the error signal (not the system output). Obviously, in the autonomous mode the singular points of $\sigma(t)$ coincide with those of $y(t)$, and the amplitudes of σ and of y are equal. Therefore, the value of the amplitude a_y determined above can be used instead of the value of the amplitude a_σ of $\sigma(t)$. Yet, at the input-output analysis with respect to

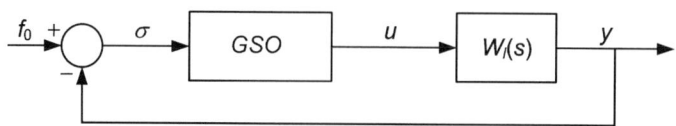

Fig. 5.3. System controlled by generalized sub-optimal algorithm

5.5 Describing function analysis of external signal propagation

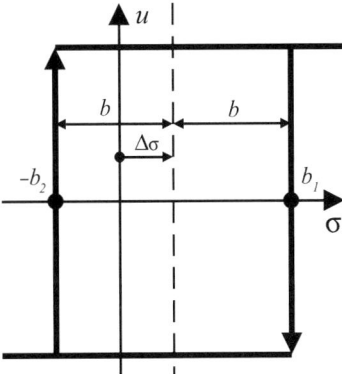

Fig. 5.4. Transformation into equivalent symmetric relay

the external constant input, the asymmetric amplitudes for positive and negative values of $\sigma(t)$ have to be considered. Because of the dependence of the hysteresis value on the amplitudes of the oscillation, the switching points of the relay (values b_1 and b_2) become asymmetric, and we have to consider the relay with asymmetric hysteresis. Moreover, the values of b_1 and b_2 are both negative due to the phase-lead character of the sub-optimal algorithm (we traditionally associate positive hysteresis with the lag in the relay characteristic).

To simplify our analysis, let us transform the original relay system into an equivalent system with the relay having a symmetric hysteresis (Fig. 5.4).

We note that if we consider an augmented error $\sigma^*(t)$

$$\sigma^*(t) = \sigma(t) - \Delta_\sigma \tag{5.11}$$

where Δ_σ is the shift of the vertical axis in Fig. 5.4 (the distance between the solid vertical axis and the dashed one) given by

$$\Delta_\sigma = \frac{b_1 - b_2}{2}, \tag{5.12}$$

then the analysis of the relay system with symmetrical hysteretic relay given by

$$b = \frac{b_1 + b_2}{2} \tag{5.13}$$

and error signal $\sigma^*(t)$ can be done using the DF method (subject to accounting for dependence of b on the amplitude of the oscillation). Therefore, we reduce the task of analysis to the symmetric one by applying transformations (5.11)–(5.13) to the system. Hence, the system can be considered as identical to the one with a symmetric hysteresis b and an additional input Δ_σ due to the error augmentation (5.11).

Now we find the dependence of Δ_σ on the value of the constant input f_0. Consider the Fourier expansion of the asymmetric periodic control $u(t)$ in Fig. 1.4,

$$u(t) = u_0 + \frac{4c}{\pi}\sum_{k=1}^{\infty}\frac{1}{k}\sin(\pi k\frac{\theta_1}{\theta_1+\theta_2})$$
$$\{\cos(k\omega\theta_1/2)\cos(k\omega t) + \sin(k\omega\theta_1/2)\sin(k\omega t)\},$$

where $u_0 = c(\theta_1 - \theta_2)/(\theta_1 + \theta_2)$, $\omega = 2\pi/(\theta_1 + \theta_2)$. Therefore, considering only the fundamental frequency component (as per the filtering hypothesis), we write

$$u(t) \approx u_0 + \frac{4c}{\pi}\sin(\pi\frac{\theta_1}{\theta_1+\theta_2}) \times \{\cos(\omega\theta_1/2)\cos(\omega t) + \sin(\omega\theta_1/2)\sin(\omega t)\}, \tag{5.14}$$

which can be re-written as follows:

$$u(t) \approx u_0 + \frac{4c}{\pi}\sin(\pi\theta_1/(\theta_1+\theta_2)) \times \cos(\omega(t - \theta_1/2)). \tag{5.15}$$

One can see from (5.15) that the amplitude of the oscillations of $u(t)$ is $\frac{4c}{\pi}\sin(\pi\theta_1/(\theta_1+\theta_2))$ (the fundamental frequency component). An application of the external constant signal f_0 results not only in the bias of the error signal $\sigma(t)$ but also in the decrease of the amplitude of the oscillation $\sigma(t)$. Moreover, b_1 and b_2 are different and depend on the positive amplitude of $\sigma(t)$ and the negative amplitude, respectively. We consider the following relationship between the hysteresis and the amplitudes of the error signal oscillation,

$$b_1 = \beta\sigma_{\min}, \quad b_2 = -\beta\sigma_{\max}$$

or

$$b_1 = \beta f_0 - \beta a_p, \quad b_2 = -\beta f_0 + \beta a_n,$$

where a_p is the "positive" amplitude of $y(t)$, $a_p > 0$, and a_n is the "negative" amplitude of $y(t)$, $a_n < 0$.

Therefore, $\Delta_\sigma = \frac{\beta}{2}(2f_0 - a_p - a_n)$ and the derivative $\frac{d\Delta_\sigma}{du_0}$ is

$$\frac{d\Delta_\sigma}{du_0} = -\frac{\beta}{2}\left(\frac{da_p}{du_0} + \frac{da_n}{du_0} - 2\frac{df_0}{du_0}\right). \tag{5.16}$$

The derivatives of (5.16) at the point corresponding to $u_0 = 0$ are found below. It follows from (5.15) that

$$a_p = u_0 + \frac{4c}{\pi}\sin(\pi\frac{\theta_1}{\theta_1+\theta_2})$$
$$a_n = u_0 - \frac{4c}{\pi}\sin(\pi\frac{\theta_1}{\theta_1+\theta_2}).$$

Also, given the relationship between the averaged control u_0 and the positive pulse duration θ_1, which is $u_0 = c(\theta_1 - \theta_2)/(\theta_1 + \theta_2) = c(2\theta_1 - T)/T$, and the derivative $\frac{du_0}{d\theta_1} = 2c/T$, we obtain formulas for the following derivatives at the point $\theta_1 = \theta_2$ ($u_0 = 0$):

$$\frac{da_p}{du_0} = \frac{da_n}{du_0} = W(j0).$$

5.5 Describing function analysis of external signal propagation   111

The derivative $\frac{df_0}{du_0}$ in formula (5.16) can be obtained from the equation of the balance of constant terms in the system

$$(f_0 - u_0 W(j0)) k_{nDF} = u_0,$$

where k_{nDF} is the equivalent gain of the sub-optimal algorithm, which we aim to find. The dependence of f_0 on u_0 is ficticious. It can be interpreted as "how much we should adjust f_0 to obtain the change in u_0 that we need." The subscript DF indicates that this is a variable derived with the describing function method.

Therefore,

$$\frac{df_0}{du_0} = \frac{1}{k_{nDF}} + W(j0)$$

and

$$\frac{d\Delta_\sigma}{du_0} = -\beta W(j0) + \beta \left(\frac{1}{k_{nDF}} + W(j0) \right) = \frac{\beta}{k_{nDF}} \quad (5.17)$$

Once the additional input due to the error augmentation is determined (formula (5.17)), we obtain an analytical formula of the equivalent gain of the sub-optimal algorithm. Let

$$k^*_{nDF} = \frac{2c}{\pi \sqrt{a^2_{yDF} - b^2}} = \frac{2c}{\pi a_{yDF} \sqrt{1-\beta^2}}$$

be the equivalent gain of the hysteretic relay. It does not account for the Δ_σ and is, therefore, not an equivalent gain of the whole algorithm. The equivalent gain of the algorithm can be determined as the equivalent gain of the relay k^*_{nDF} having a feedback with the gain $\frac{\beta}{k_{nDF}}$:

$$k_{nDF} = \frac{k^*_{nDF}}{1 - \beta k^*_{nDF}/k_{nDF}}.$$

We find from the previous equation that

$$k_{nDF} = k^*_{nDF} (1 + \beta). \quad (5.18)$$

The rest of the analysis is the same as for the conventional relay system. The effect of the error augmentation via the gain (5.18) can be depicted as in Fig. 5.5.

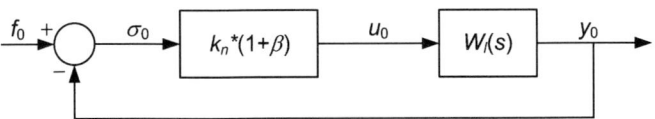

Fig. 5.5. Equivalent system for averaged components propagation through system with SOSM

The results obtained for the constant input f_0 can be extended to the case of slowly varying f_0, where "slow" implies the fact that the frequency of the input signal is much lower than the frequency of chattering. In Fig. 5.5, the results obtained for the constant input f_0 are applied to slow varying input $f(t)$. As a result, the plant gain $W(j0)$ is replaced with the plant gain at the frequency of the input signal. One can see that application of the advance switching as per the sub-optimal SOSM algorithm results in an increase of the equivalent gain of the relay. This increase depends on the parameter β of the sub-optimal algorithm.

5.6 Exact frequency-domain analysis of external signal propagation

We carry out the same input-output analysis as above but via the LPRS method. Consider the system controlled by the sub-optimal algorithm, to which an external constant input f_0 is applied as in Fig. 5.3.

Because of the dependence of the hysteresis value on the amplitudes of the oscillation, the switching points of the relay (values b_1 and b_2) become asymmetric, and we have to consider the relay with asymmetric hysteresis. As before, to simplify our analysis by utilizing the LPRS method, let us transform the original relay system into an equivalent system with the relay having a symmetric hysteresis (Fig. 5.4). We note that if we consider an augmented error $\sigma^*(t)$, with Δ_σ being the shift of the vertical axis in Fig. 5.4 (the distance between the solid vertical axis and the dashed one), then analysis of the relay system with symmetrical hysteretic relay can be done with the LPRS method (subject to accounting for dependence of b on the amplitude of the oscillation). Therefore, we can reduce the task of analysis to the symmetric one by applying transformations (5.11)–(5.13) to the system.

Now consider the dependence of the average error σ_0 and the average control u_0 on the constant input f_0. Simulations show that these two dependences are close to linear in a wide range (see example below). The dependence of u_0 on σ_0 (due to the variation of f_0) will be referred to as the *bias function*. Since these two dependences are linear, the *bias function* is linear too. Therefore, we can approximate the *bias function* with the *equivalent gain* and carry out an analysis of propagation of averaged (on the period of chattering) variables through the sliding mode system as an analysis of a linear system. The *equivalent gain* is determined as the slope of the *bias function*. We derive an expression for the *equivalent gain*.

We use the following approach. Consider the unequally spaced control $u(t)$ of amplitude c with the positive and negative pulse durations being θ_1 and θ_2, (respectively, in the same way as it was done in Chapter 2). After that, we analyze the response of the linear part to this control and obtain the value of constant input f_0 and the relay hysteresis that is needed to produce the given

5.6 Exact frequency-domain analysis of external signal propagation

$u(t)$ in the closed-loop system. To find the equivalent gain, we find the reciprocal derivative:

$$\frac{d\sigma_0}{du_0} = \frac{1}{k_n}. \tag{5.19}$$

It follows from (5.11) that

$$\frac{d\sigma_0}{du_0} = \frac{d\sigma_0^*}{du_0} + \frac{d\Delta_\sigma}{du_0} = \frac{1}{k_{nLPRS}} + \frac{d\Delta_\sigma}{du_0} \tag{5.20}$$

where k_{nLPRS} is the equivalent gain obtained via the LPRS for the equivalent system with symmetric hysteresis $2b$ (see Chapter 2). The second component in formula (5.20) is given by formula (5.16). The last derivative in (5.16) can be written as follows (the same as in the DF analysis),

$$\frac{df_0}{du_0} = \frac{1}{k_n} + W(j0). \tag{5.21}$$

The first two derivatives in (5.16) are found below.
The output of the linear part is

$$\begin{aligned}y(t) = y_0 + \tfrac{4c}{\pi} \sum_{k=1}^{\infty} & \sin(\pi k\theta_1/(\theta_1+\theta_2))/k \\ \times & \{\cos(k\omega\theta_1/2)\cos[k\omega t + \varphi_L(k\omega)] \\ & + \sin(k\omega\theta_1/2)\sin[k\omega t + \varphi_L(k\omega)]\} A_L(k\omega).\end{aligned} \tag{5.22}$$

We can rewrite the last formula (given that $\theta_1 + \theta_2 = T = \frac{2\pi}{\omega}$) as follows:

$$\begin{aligned}y(t) = y_0 + \tfrac{4c}{\pi} \sum_{k=1}^{\infty} & \tfrac{1}{k}\sin(\tfrac{k}{2}\omega\theta_1) \\ \times & \{\cos(\tfrac{k}{2}\omega\theta_1)\cos[k\omega t + \varphi_L(k\omega)] \\ & + \sin(\tfrac{k}{2}\omega\theta_1)\sin[k\omega t + \varphi_L(k\omega)]\} A_L(k\omega).\end{aligned} \tag{5.23}$$

We find the derivative of $y(t)$ with respect to θ_1:

$$\begin{aligned}\tfrac{\partial y(t)}{\partial \theta_1} = \tfrac{\partial y_0}{\partial \theta_1} + \tfrac{4c}{\pi} \sum_{k=1}^{\infty} \tfrac{1}{k} & \{\tfrac{k}{2}\omega \, \cos(\tfrac{k}{2}\omega\theta_1) \times \{\cos(\tfrac{k}{2}\omega\theta_1)\cos[k\omega t + \varphi_L(k\omega)] \\ & + \sin(\tfrac{k}{2}\omega\theta_1)\sin[k\omega t + \varphi_L(k\omega)]\} A_L(k\omega) \\ & + \sin(\tfrac{k}{2}\omega\theta_1) \times \tfrac{k}{2}\omega \, \{-\sin(\tfrac{k}{2}\omega\theta_1)\cos[k\omega t + \varphi_L(k\omega)] \\ & + \cos(\tfrac{k}{2}\omega\theta_1)\sin[k\omega t + \varphi_L(k\omega)]\} A_L(k\omega)\}.\end{aligned} \tag{5.24}$$

At $\theta_1 = \pi/\omega$, this derivative becomes

$$\frac{\partial y(t)}{\partial \theta_1} = \frac{\partial y_0}{\partial \theta_1} + \frac{4c}{\pi} \sum_{k=1}^{\infty} \frac{1}{k} \left\{ \frac{k}{2}\omega \, \cos(\frac{k\pi}{2}) \times \{\cos(\frac{k\pi}{2})\cos[k\omega t + \varphi_L(k\omega)] \right.$$
$$+ \sin(\frac{k\pi}{2}) \sin[k\omega t + \varphi_L(k\omega)]\} A_L(k\omega)$$
$$+ \sin(\frac{k\pi}{2}) \times \frac{k}{2}\omega \, \{-\sin(\frac{k\pi}{2})\cos[k\omega t + \varphi_L(k\omega)]$$
$$\left. + \cos(\frac{k\pi}{2}) \sin[k\omega t + \varphi_L(k\omega)]\} A_L(k\omega) \right\}$$
(5.25)

or

$$\frac{\partial y(t)}{\partial \theta_1} = \frac{\partial y_0}{\partial \theta_1} + \frac{4c}{\pi} \sum_{k=1}^{\infty} \frac{1}{k} \{ \frac{k}{2}\omega \, \cos^2(\frac{k\pi}{2}) \cos[k\omega t + \varphi_L(k\omega)] \, A_L(k\omega)$$
$$- \frac{k}{2}\omega \, \sin^2(\frac{k\pi}{2}) \cos[k\omega t + \varphi_L(k\omega)] A_L(k\omega) \}.$$
(5.26)

Formula (5.26) can be transformed into:

$$\frac{\partial y(t)}{\partial \theta_1} = \frac{\partial y_0}{\partial \theta_1} + \frac{2c}{\pi} \sum_{k=1}^{\infty} \omega \left[\cos^2(\frac{k\pi}{2}) - \sin^2(\frac{k\pi}{2}) \right] \cos[k\omega t + \varphi_L(k\omega)] A_L(k\omega),$$
(5.27)

which leads to the following formula when we expand the *cosine* of a sum:

$$\frac{\partial y(t)}{\partial \theta_1} = \frac{\partial y_0}{\partial \theta_1} + \frac{2c}{\pi} \sum_{k=1}^{\infty} \omega \left[\cos^2(\frac{k\pi}{2}) - \sin^2(\frac{k\pi}{2}) \right] [\cos(k\omega t) \cos \varphi_L(k\omega)$$
$$- \sin(k\omega t) \sin \varphi_L(k\omega)] A_L(k\omega).$$
(5.28)

For even k, the term in $[\ldots]$ is 1, and for odd k it is -1. Therefore,

$$\frac{\partial y(t)}{\partial \theta_1} = \frac{\partial y_0}{\partial \theta_1} + \frac{2c}{\pi} \sum_{k=1}^{\infty} \omega \, (-1)^k \, [\cos(k\omega t) \cos \varphi_L(k\omega)$$
$$- \sin(k\omega t) \sin \varphi_L(k\omega)] A_L(k\omega). \quad (5.29)$$

Let us compare the derivative values for time $t = t_m$ and time $t = \pi/\omega + t_m$, where t_m is the time of maximum of $y(t)$. It follows from (5.29) that

$$\left.\frac{\partial y(t)}{\partial \theta_1}\right|_{t=\pi/\omega+t_m} = \frac{\partial y_0}{\partial \theta_1} + \frac{2c}{\pi} \sum_{k=1}^{\infty} \omega \, (-1)^k \, [\cos(k\omega(t_m + \pi/\omega)) \cos \varphi_L(k\omega)$$
$$- \sin(k\omega(t_m + \pi/\omega)) \sin \varphi_L(k\omega)] A_L(k\omega)$$
(5.30)

and after applying the sum formulas for *sine* and *cosine*:

$$\left.\frac{\partial y(t)}{\partial \theta_1}\right|_{t=\pi/\omega+t_m} = \frac{\partial y_0}{\partial \theta_1} + \frac{2c}{\pi} \sum_{k=1}^{\infty} \omega \, (-1)^k \, [(\cos(k\omega t_m) \cos(k\pi) - \sin(k\omega t_m) \sin(k\pi))$$
$$\times \cos \varphi_L(k\omega) - (\sin(k\omega t_m) \cos(k\pi)$$
$$+ \cos(k\omega t_m) \sin(k\pi)) \sin \varphi_L(k\omega)] A_L(k\omega).$$
(5.31)

5.6 Exact frequency-domain analysis of external signal propagation

For every k (odd and even), the following equality holds: $\sin(k\pi) = 0$. Therefore, (5.31) can be reduced to

$$\frac{\partial y(t)}{\partial \theta_1}\bigg|_{t=\pi/\omega+t_m} = \frac{\partial y_0}{\partial \theta_1} + \frac{2c}{\pi}\sum_{k=1}^{\infty}\omega\,(-1)^k\,[\cos(k\omega t_m)\cos(k\pi)\cos\varphi_L(k\omega)$$
$$- \sin(k\omega t_m)\cos(k\pi)\sin\varphi_L(k\omega)]A_L(k\omega) \quad (5.32)$$

or

$$\frac{\partial y(t)}{\partial \theta_1}\bigg|_{t=\pi/\omega+t_m} = \frac{\partial y_0}{\partial \theta_1} + \frac{2c}{\pi}\sum_{k=1}^{\infty}\omega\,(-1)^k\cos(k\pi)\,[\cos(k\omega t_m)\cos\varphi_L(k\omega)$$
$$- \sin(k\omega t_m)\sin\varphi_L(k\omega)]A_L(k\omega)$$

$$= \frac{\partial y_0}{\partial \theta_1} + \frac{2c}{\pi}\sum_{k=1}^{\infty}\omega\,(-1)^k(-1)^k\,[\cos(k\omega t_m)\cos\varphi_L(k\omega)$$
$$- \sin(k\omega t_m)\sin\varphi_L(k\omega)]A_L(k\omega)$$

$$= \frac{\partial y_0}{\partial \theta_1} + \frac{2c}{\pi}\sum_{k=1}^{\infty}\omega\,[\cos(k\omega t_m)\cos\varphi_L(k\omega)$$
$$- \sin(k\omega t_m)\sin\varphi_L(k\omega)]A_L(k\omega). \quad (5.33)$$

A similar derivative for time $t = t_m$ can be directly obtained from (5.29):

$$\frac{\partial y(t)}{\partial \theta_1}\bigg|_{t=t_m} = \frac{\partial y_0}{\partial \theta_1} + \frac{2c}{\pi}\sum_{k=1}^{\infty}\omega(-1)^k\,[\cos(k\omega t_m)\cos\varphi_L(k\omega)$$
$$- \sin(k\omega t_m)\sin\varphi_L(k\omega)]A_L(k\omega). \quad (5.34)$$

Comparing (5.33) and (5.34), we see that positive and negative parts of the plant output signal respond at different rates to the changes in the pulse width (t_m is considered an arbitrary value).

The derivatives of the positive and the negative amplitudes are determined as follows:

$$\frac{da_p}{d\theta_1} = \frac{\partial y(t)}{\partial \theta_1}\bigg|_{t=t_m} + \frac{\partial y(t)}{\partial t}\bigg|_{t=t_m}\frac{dt_m}{d\theta_1}.$$

Given that $\frac{\partial y(t)}{\partial t}\bigg|_{t=t_m} = 0$ (where t_m is both the time of maximum and the time of minimum), the following holds. For the "positive" amplitude

$$\frac{\partial a_p}{\partial \theta_1} = \frac{\partial y_0}{\partial \theta_1} + \frac{2c}{\pi}\omega\sum_{k=1}^{\infty}(-1)^k\,[\cos(k\omega t_{\max})\cos\varphi_L(k\omega)$$
$$- \sin(k\omega t_{\max})\sin\varphi_L(k\omega)]A_L(k\omega) \quad (5.35)$$

and for the "negative" amplitude

$$\frac{\partial a_n}{\partial \theta_1} = \frac{\partial y_0}{\partial \theta_1} + \frac{2c}{\pi}\omega\sum_{k=1}^{\infty}[\cos(k\omega t_{\max})\cos\varphi_L(k\omega)$$
$$- \sin(k\omega t_{\max})\sin\varphi_L(k\omega)]A_L(k\omega). \quad (5.36)$$

In (5.35) and (5.36), the time instant is the same: $t_m = t_{max}$, $a_n = y(t_{min})$. From formula (5.18), we obtain the derivative for the shift

$$\frac{\partial \Delta_\sigma}{\partial \theta_1} = -0.5\beta \left(\frac{\partial a_p}{\partial \theta_1} + \frac{\partial a_n}{\partial \theta_1} - 2\frac{df_0}{d\theta_1} \right), \tag{5.37}$$

where the derivatives in parentheses are given by (5.35) and (5.36).

We compute this derivative:

$$\frac{\partial \Delta_\sigma}{\partial \theta_1} = -\beta \left\{ -\frac{df_0}{d\theta_1} + \frac{\partial y_0}{\partial \theta_1} + \frac{2c}{\pi}\omega \sum_{k=1}^{\infty} [\cos(2k\omega t_{max}) \cos \varphi_L(2k\omega) \right.$$
$$\left. - \sin(2k\omega t_{max}) \sin \varphi_L(2k\omega)] A_L(2k\omega) \right\}$$
$$= -\beta \left\{ -\frac{df_0}{d\theta_1} + \frac{\partial y_0}{\partial \theta_1} + \frac{2c}{\pi}\omega \sum_{k=1}^{\infty} [\cos(2k\omega t_{max} + \varphi_L(2k\omega))] A_L(2k\omega) \right\}. \tag{5.38}$$

Now we compute the derivative with respect to u_0. Given that $u_0 = c\frac{\theta_1 - \theta_2}{\theta_1 + \theta_2}$ and that the frequency is constant, the following holds: $u_0 = c\left(\frac{\theta_1 \omega}{\pi} - 1\right)$. Therefore, the derivative is $\frac{d\theta_1}{du_0} = \frac{\pi}{c\omega}$. Also, it is obvious that $\frac{dy_0}{du_0} = W_l(j0)$, and as a result:

$$\frac{\partial \Delta_\sigma}{\partial u_0} = -\beta \left\{ -\frac{df_0}{du_0} + W_l(j0) + 2\sum_{k=1}^{\infty} [\cos(2k\omega t_{max} + \varphi_L(2k\omega))] A_L(2k\omega) \right\}. \tag{5.39}$$

Formula (5.39) includes the component $\frac{df_0}{du_0}$ that can only be determined if the equivalent gain of the sub-optimal algorithm is known. Yet, it is also the variable which we aim to determine. Consequently, computing the derivative (5.39) involves solving an equation for k_n. This component of (5.39) is given as follows:

$$\frac{df_0}{du_0} = \frac{1}{k_n} + W_l(j0).$$

Denote the series on the right-hand side of formula (5.39) multiplied by a factor of -0.5 as $R(\omega)$, where

$$R(\omega) = \sum_{k=1}^{\infty} [\cos(2k\omega t_{max} + \varphi_L(2k\omega))] A_L(2k\omega). \tag{5.40}$$

The function $R(\omega)$ accounts for the unequal response of the negative and positive amplitudes to the change of u_0. It contains only even harmonics of the fundamental frequency component. It is a function of the linear plant parameters and can be computed from the transfer function of the plant. Therefore, (5.39) can be rewritten as follows,

$$\frac{\partial \Delta_\sigma}{\partial u_0} = \frac{\beta}{k_n} - 2\beta R(\Omega) \tag{5.41}$$

where Ω is the frequency of chattering. The equivalent gain of the hysteretic relay computed as per the LPRS method is: $k^*_{nLPRS} = \frac{1}{-2\mathrm{Re}J(\Omega)}$, where $J(\omega)$ is the LPRS. Now we write an equation from which the formula of the equivalent gain of the sub-optimal algorithm can be found. This equation considers the effect of the shift of the asymmetric hysteretic relay characteristic by Δ_σ as the equivalent gain of the relay k^*_{nLPRS} having a feedback with the gain $\frac{\partial \Delta_\sigma}{\partial u_0}$ (formula (5.41)):

$$k_n = \frac{k^*_{nLPRS}}{1 - k^*_{nLPRS}\beta\left(\frac{1}{k_n} - 2R(\Omega)\right)}. \tag{5.42}$$

Equation (5.42) can be solved for k_n, and the formula of the equivalent gain of the sub-optimal algorithm is expressed as:

$$k_n = \frac{(1+\beta)\,k^*_{nLPRS}}{1 + 2\beta k^*_{nLPRS}R(\Omega)} \tag{5.43}$$

Presumably, the term in the denominator of (5.43) $2\beta R(\Omega)$ is small (if the oscillation is harmonic, it is zero). In this case, the value of the equivalent gain computed as in (5.43) is approximately equal to the equivalent gain value computed as in (5.18) (the difference is due to the difference between the values of k^*_{nLPRS} and k^*_{nDF}). Formula (5.43) can be rewritten with LPRS notation:

$$k_n = \frac{1+\beta}{-2\mathrm{Re}J(\Omega) + 2\beta R(\Omega)}. \tag{5.44}$$

Therefore, because $\mathrm{Re}J(\Omega)$ is normally a negative value, the plot $-2\mathrm{Re}J(\Omega) + 2\beta R(\Omega)$ is depicted as a small offset of the LPRS in either positive or negative direction. Yet the main change of the equivalent gain value (in comparison with the relay control) is due to the multiplier $(1+\beta)$.

Note: it was assumed above that $\left.\frac{\partial \Omega}{\partial u_0}\right|_{u_0=0} = 0$. This property follows from the symmetry principle: both positive and negative changes of u_0 around $u_0 = 0$ result in the same changes of the oscillation frequency Ω. Hence, the derivative of the frequency should be zero: $\left.\frac{\partial \Omega}{\partial u_0}\right|_{u_0=0} = 0$. The proof of this property for conventional relay control was given above.

5.7 Example of the analysis of sub-optimal algorithm performance

We carry out the DF and LPRS analysis of a plant controlled by the sub-optimal algorithm. Consider $W(s)$ being the cascade connection of the second-order linear plant $W_p(s)$ and the first-order plus dead-time dynamic actuator $W_a(s)$

Table 5.1. Analysis of periodic motions for the example of Section 5.7

	Frequency [rad s^{-1}]	Amplitude
DF	3.887	0.0867
Exact frequency-domain analysis	3.743	0.0968
Simulation	3.705	0.0934

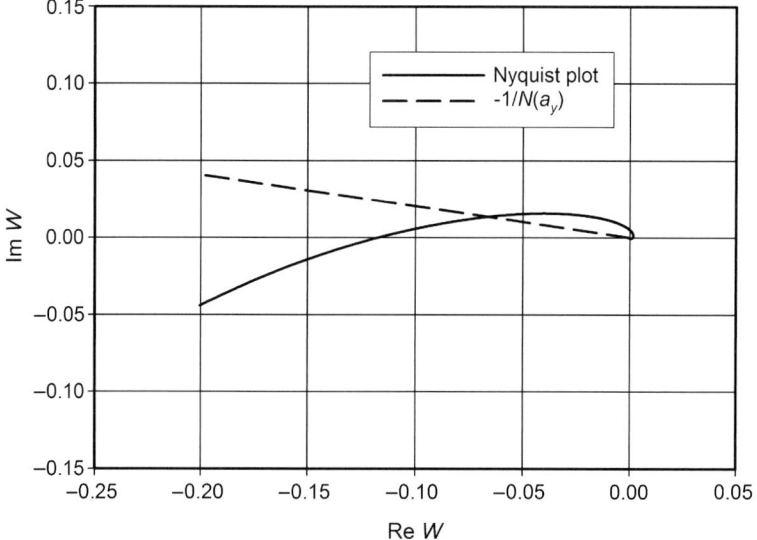

Fig. 5.6. DF analysis

$$W_p(s) = \frac{1}{s^2 + s + 1}, \quad W_a(s) = \frac{e^{-0.1s}}{0.02s + 1}. \quad (5.45)$$

The loop is closed via the sub-optimal algorithm (5.2) with switching anticipation parameter $\beta = 0.2$ and control magnitude $c = 1$. The approximate and theoretically exact parameters of the periodic solution [obtained by means of the DF and the exact method via application of equation (5.7)] are computed and found through computer simulations and presented in Table 5.1. A higher accuracy of the exact frequency-domain approach is apparent.

The plots for the DF analysis are given in Fig. 5.6. The magnitude of $W(j\omega)$ at the intersection point is $M = |W(j3.887)| \approx 0.0681$, and the estimated oscillation amplitude is $a_{yDF} = 4Mc/\pi \approx 0.0867$.

Plots for the DF analysis are given in Fig. 5.6. The two plots given in Fig. 5.7 refer to the exact frequency-domain analysis. The upper plot shows the function $\Phi(\omega)$ drawn in the frequency interval $\omega \in [1.3; 30]$rad/s, and the bottom plot focuses on the frequency range $\omega \in [2.9; 30]$rad/s, where the

5.7 Example of the analysis of sub-optimal algorithm performance 119

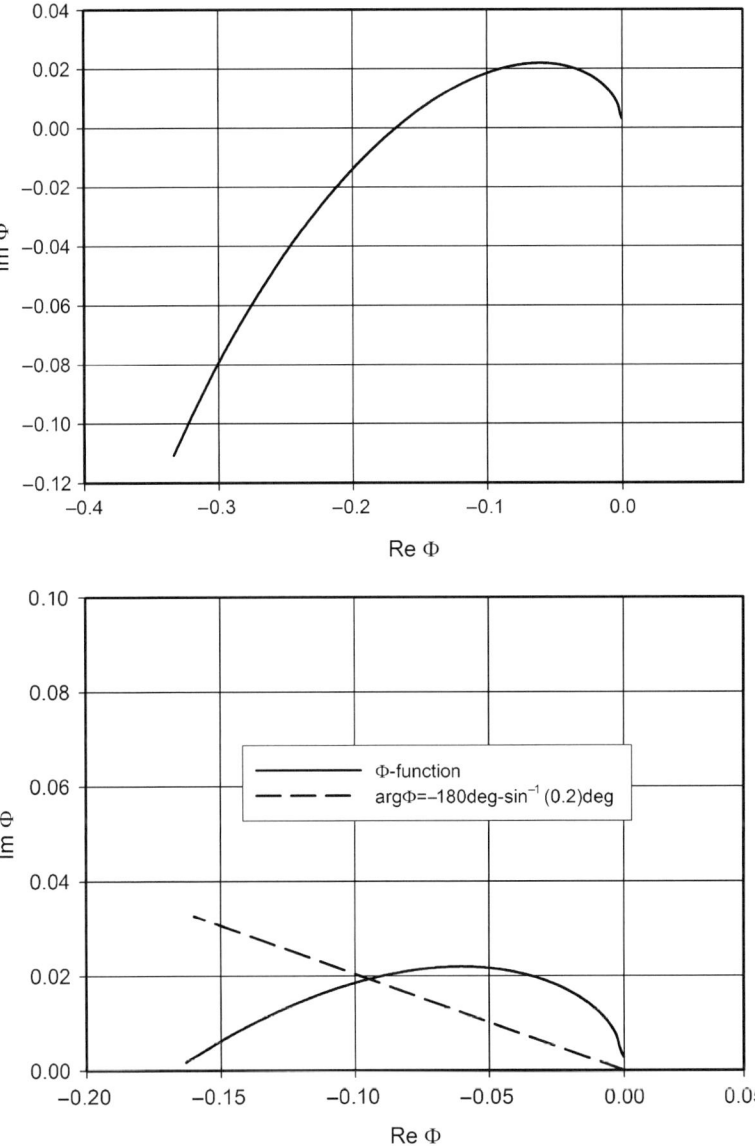

Fig. 5.7. LPRS analysis

intersection with the negative reciprocal DF is found. Figure 5.8 provides the results of the computer simulation of this system.

The higher accuracy of the exact frequency-domain analysis, with respect to the DF analysis, is due to the actual shape of the oscillations and the use of the true amplitude versus the amplitude of the fundamental frequency component. The mismatch between the simulation values and those computed

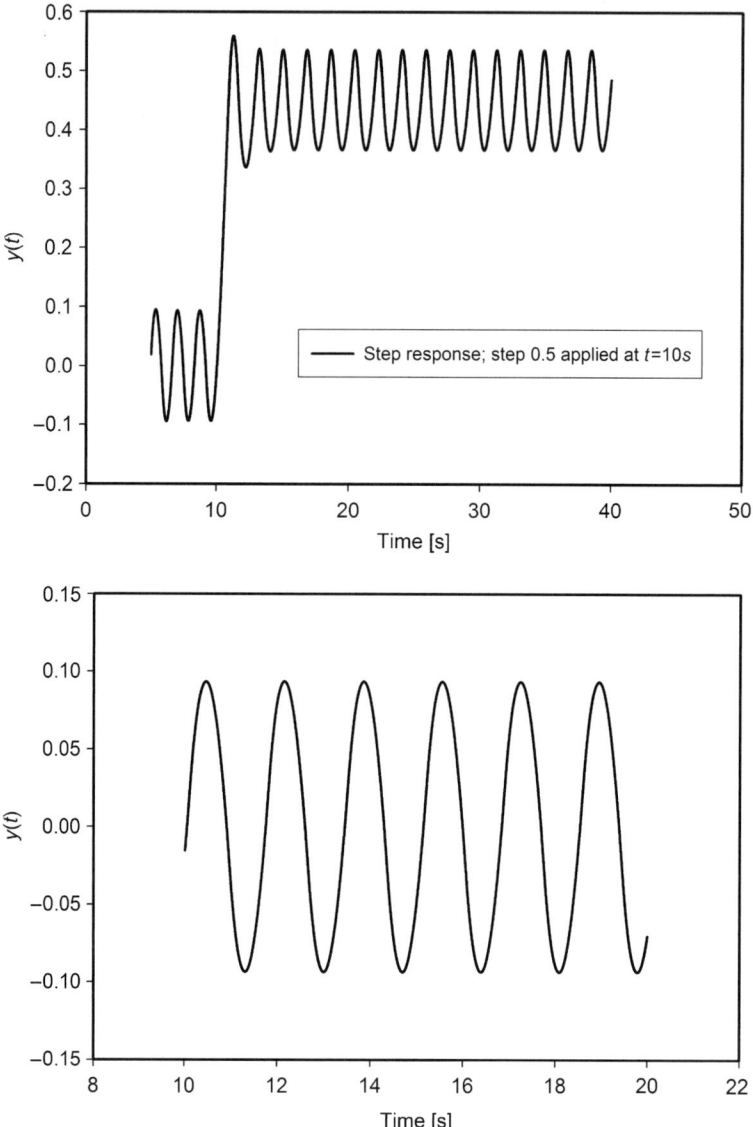

Fig. 5.8. Periodic motion in the system controlled by the sub-optimal algorithm (simulations)

via the exact frequency-domain analysis is caused by the factors of numerical approximation such as the truncation of the series (5.9), the round-offs, and the discrete-time integration of the simulation example.

Now we continue with the input-output analysis and find the equivalent gain value as in the DF method:

5.7 Example of the analysis of sub-optimal algorithm performance

$$k_{nDF}^* = \frac{2c}{\pi\sqrt{a_{yDF}^2 - b^2}} = \frac{2c}{\pi a_{yDF}\sqrt{1-\beta^2}} \approx 5.83.$$

Therefore, the transfer function of the closed-loop system with the sub-optimal algorithm is as follows:

$$W(s) = \frac{k_{nDF}^*(1+\beta)W_p(s)W_a(s)}{1 + k_{nDF}^*(1+\beta)W_p(s)W_a(s)} \quad (5.46)$$

and the equivalent gain of the sub-optimal algorithm is

$$k_n = k_{nDF}^*(1+\beta) = 5.83 \cdot 1.2 = 7.00.$$

The input-output LPRS analysis gives a very close value of the equivalent gain to the one obtained via the DF method. This happens due to very good low-pass filtering properties of the plant (see simulations below).

The results obtained analytically via the DF method are verified below via simulations. The experimental values of the frequency and amplitude are given in Table 5.1. The bias function measured in the simulations is given in Fig. 5.9. There is some oscillatory component in the experimentally measured bias function. For that reason, the equivalent gain can be obtained as a linear regression of the dependence given in Fig. 5.9. The experimental value of the equivalent gain is $k_{nSIM} = 7.03$. Therefore, both methods of analysis provide a good estimate of the input-output properties of the system controlled by the sub-optimal algorithm. As a result, the undertaken simulation via the measurement of the equivalent gain serves as a validation of the formula of the transfer

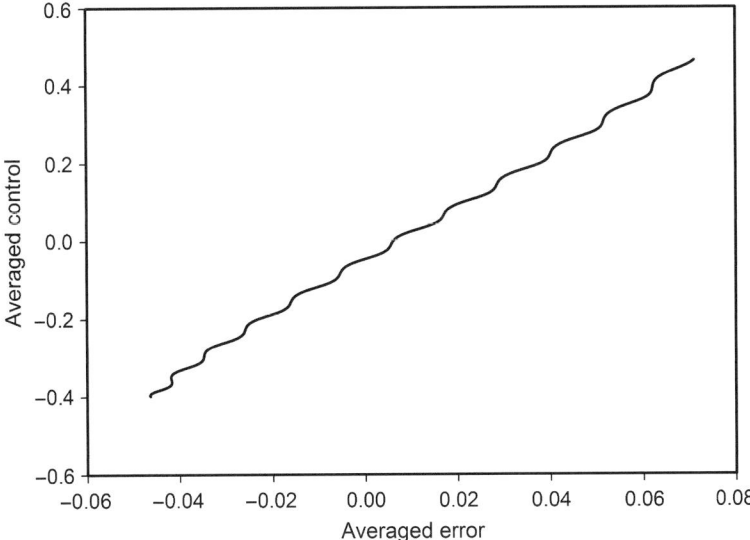

Fig. 5.9. Bias function obtained via simulations

function of the closed-loop system (5.43). Moreover, the whole concept of the equivalent gain is justified through simulations, and the value of the equivalent gain is verified.

5.8 Conclusions

We analyze the second-order sliding mode control algorithm *sub-optimal* in the frequency domain with the use of the DF and the LPRS methods. We demonstrate that the nature of the SOSM in comparison with the conventional first-order SM is due to the *advance switching* of the relay, which results in the finite-time convergence of the transient process if the relative degree of the plant dynamics is two. However, due to the inevitable existence of parasitic dynamics, the relative degree of the actuator-plant-sensor dynamics can never be *two*. It is always higher. As a result, chattering in the form of high-frequency oscillations occurs in systems controlled by a SOSM. The frequency of those oscillations is always higher than the frequency of the oscillations in the corresponding system with the relay controller, though.

Also, the input-output properties of the SOSM are analyzed via the DF and the LPRS methods. We show that the application of the SOSM results in an increase of the equivalent gain of the discontinuous nonlinearity (relay) for the averaged variables. This increase depends on the "anticipation" parameter β of the *sub-optimal* algorithm. We analyze the relationships between the "anticipation" parameter and the frequency and the amplitude of chattering, as well as its effect on the equivalent gain of the relay. The example given in Section 5.7 demonstrates the frequency-domain methodology of analysis and justifies the conclusions.

Part II

Applications of the locus of a perturbed relay system

6
Relay pneumatic servomechanism design

6.1 Relay pneumatic servomechanism dynamics and characteristics

Relay servomechanisms were first designed in the 1940s [17, 71] and are still being widely used now in aerospace applications. Among their advantages over linear servomechanisms are simplicity, low cost, reliability, lower weight, and, as a rule, better dynamical characteristics [56]. The basic design of a pneumatic servomechanism is given schematically in Fig. 6.1.

The main components of a servomechanism are a source of the pressurized air, an air valve, a solenoid, an air cylinder, a piston position sensor, a comparison device, a compensating filter, and a relay (class D) amplifier. The servomechanism operates as follows. Pressurized air enters into either the left or the right chambers of the air cylinder, depending on the air valve position. Because of the pressure differential in the cylinder chambers, the piston moves in the appropriate direction. The position sensor measures the position of the cylinder and sends a signal proportional to the piston displacement from the neutral position to the comparison device, which provides an error signal equal to the difference between the reference input U_{inp} and the sensor signal. If the error is positive, the relay energizes the solenoid, which moves the air valve (to the left, as in Fig. 6.1); if the error is negative, the solenoid is deenergized and the spring moves the air valve to the right. There are some other designs of the relay pneumatic servomechanism: with two solenoids, or the solenoid with two different windings, different types of air valves, etc.

By modulating the flow of pressurized air into the air cylinder (this is the task of the control devices), one can regulate the piston position. If the reference input to the servomechanism is zero, a symmetric periodic motion occurs in the servomechanism loop; the piston has small oscillations around the zero position, the position measurements (through the comparator) are transformed into pulses and amplified by the relay amplifier, the pulses drive the solenoid and the air valve, the air is alternately let into the first or the second

6 Relay pneumatic servomechanism design

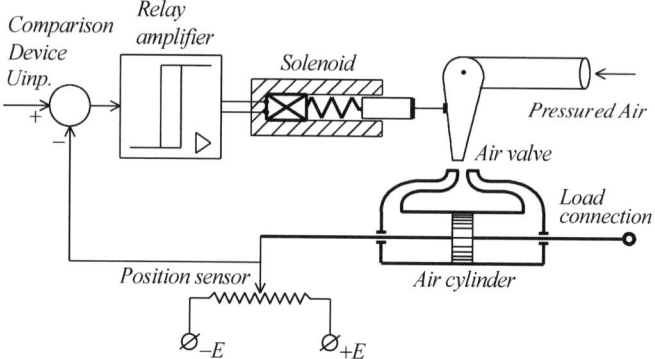

Fig. 6.1. Relay pneumatic servomechanism

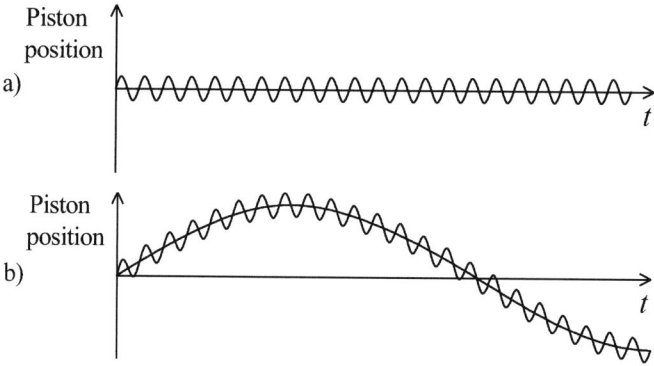

Fig. 6.2. Piston motion: (a) in the autonomous mode and (b) under external excitation

chambers of the cylinder, and the piston moves back and forth because of the pressure differential. A self-sustained oscillation occurs as a result (Fig. 6.2a). The frequency of these oscillations for small-size pneumatic servomechanisms can reach over 100 Hz. The higher frequency of self-excited oscillations reveals the faster dynamics of the servomechanism.

If a slow input signal is applied to the servomechanism, it modulates the self-sustained oscillations, and the resulting motion of the piston is a combination of self-sustained oscillations and forced motion (Fig. 6.2b).

The model of the electro-pneumatic servomechanism built from first principles (the laws of thermodynamics) is nonlinear. In practice, the linearized model is commonly used. It is presented as a block diagram in Fig. 6.3. The compensator is not shown in this diagram, and it thus represents the model of the uncompensated servomechanism.

In this diagram, f is an external excitation, u is the relay amplifier output, α is the position of the solenoid, F is the force developed by the pressure

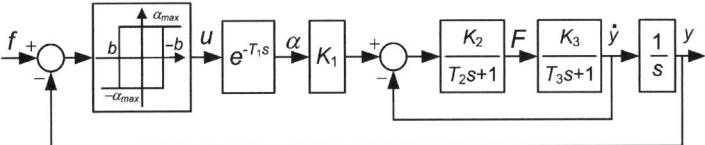

Fig. 6.3. Uncompensated electro-pneumatic servomechanism

differential and applied to the piston, $\dot{y}$ is the velocity of the piston, y is the piston position (displacement from the neutral position), K_i and T_i, $i = \overline{1,3}$ are gains and time constants, respectively, $2b$ is the relay hysteresis, and $\alpha_{\max}$ is the maximal stroke of the solenoid (the relay output is calibrated in the values of the solenoid stroke).

Usually, servomechanism performance is characterized by the magnitude frequency response and the phase frequency response in a certain range of frequencies and amplitudes of the harmonic excitation. The requirements are defined as an envelope within which the magnitude and the phase responses of the servomechanism should be located. Normally, the phase response is a declining function of the frequency, and instead of the envelope specifications, the maximum phase lag at the highest frequency of the external excitation is specified. Two other characteristics are the frequency and the amplitude (of the piston strokes) of the self-sustained (self-excited) oscillation in the servomechanism loop. Usually, the lower boundary of the frequency and the upper boundary of the amplitude of self-excited oscillations are specified. These two characteristics are related to each other. In fact, if the parameters of the servomechanism are constant, one of the two characteristics can be obtained from the other. However, if the parameters of the servomechanism vary, the use of both characteristics is necessary.

6.2 LPRS analysis of uncompensated relay electro-pneumatic servomechanism

The overall design of an electro-pneumatic servomechanism includes the steps of generation of the specification for the servomechanism components, selection or design of those components, analysis of dynamics of the uncompensated servomechanism, and design of compensating filters. The first two steps are beyond the scope of this book. We assume that all components of the servomechanism are properly designed or selected and, if connected with each other in the loop given by Fig. 6.1, make a properly working system. By "properly working," we mean that self-excited oscillations exist and under external excitation the system exhibits tracking of this external signal in the sense illustrated by Fig. 6.2. However, we assume that the requirements for the system characteristics specified above may not be satisfied. We call this system an uncompensated electro-pneumatic servomechanism. The analytic methodology logically follows from the LPRS method, the plant model, and the specified characteristics of the relay pneumatic servomechanism.

First, the dynamic characteristics of the uncompensated servomechanism must be computed and checked to ensure the specifications are satisfied. After that, the compensating filters must be designed to meet the required specifications. If we consider an unloaded servomechanism, which is depicted in Fig. 6.3, the linear part of the relay servo system is an integrating plant. For that reason, the LPRS formula for integrating linear parts (2.32) or formula (2.19) must be used for the calculations as follows:

$$J(\omega) = \sum_{k=1}^{\infty}(-1)^{k+1}\mathrm{Re}W_l(k\omega) + j\sum_{k=1}^{\infty}\mathrm{Im}W_l[(2k-1)\omega]/(2k-1) \quad (6.1)$$

where

$$W_l(s) = \frac{K_1 K_2 K_3}{1 + K_2 K_3} \frac{e^{-T_1 s}}{s\left(\frac{T_2 T_3}{1+K_2 K_3}s^2 + \frac{T_2+T_3}{1+K_2 K_3}s + 1\right)}. \quad (6.2)$$

The frequency of the self-sustained periodic motion is calculated by solving the equation

$$\mathrm{Im}J(\Omega) = -\frac{\pi b}{4c}, \quad (6.3)$$

and the equivalent gain k_n is calculated as per (2.4) as a function of the frequency Ω computed above.

Next, the transfer function of the closed-loop uncompensated relay servomechanism is evaluated as

$$W_{closed}(s) = \frac{k_n W_l(s)}{1 + k_n W_l(s)}, \quad (6.4)$$

and the magnitude frequency response and the phase frequency response are found using the following formulas, respectively:

$$M(\omega) = 20 \lg |W_{closed}(j\omega)| \quad (6.5)$$

$$\varphi(\omega) = \arg W_{closed}(j\omega). \quad (6.6)$$

6.3 Compensator design in the relay electro-pneumatic servomechanism

Relay control involves modes in which the frequency of self-excited oscillations (switching frequency) is typically much higher than the closed-loop system bandwidth (that corresponds with the frequency range of the external input signals). Therefore, the equivalent gain of the relay in the closed-loop system and input-output properties of the plant are defined by different frequency

6.3 Compensator design in the relay electro-pneumatic servomechanism 129

ranges of the LPRS. The input-output properties of the relay (the equivalent gain) are defined by the shape of the LPRS at frequencies near the switching frequency, whereas the input-output properties of the plant depend on the characteristics of the plant at frequencies within the system bandwidth. In order to have a larger equivalent gain, the point of intersection with the real axis must be closer to the origin. At the same time, the LPRS shape in the frequency range of the system bandwidth must be preserved. Therefore, the idea of the compensating filters design is based on the change of the shape of the LPRS at frequencies near the frequency of self-excited oscillations, without affecting the LPRS values at the frequencies of the input signals, which would result in an increase of the equivalent gain of the relay and enhancement of the closed-loop performance of the servomechanism. The LPRS gives a very demonstrative illustration of this idea (Fig. 6.4).

Therefore, compensation can be defined as the selection of linear filters and their connection within the system so that the LPRS of the system has a desired location at the frequencies near the switching (chattering) frequency and does not change the open-loop characteristics of the system at frequencies within the specified bandwidth.

The desired configuration of the LPRS can be evaluated from the specification on the system closed-loop performance in the specified bandwidth (provided that the performance enhancement is achieved only by an increase of the *equivalent gain* k_n, so that the frequency properties of the open-loop sys-

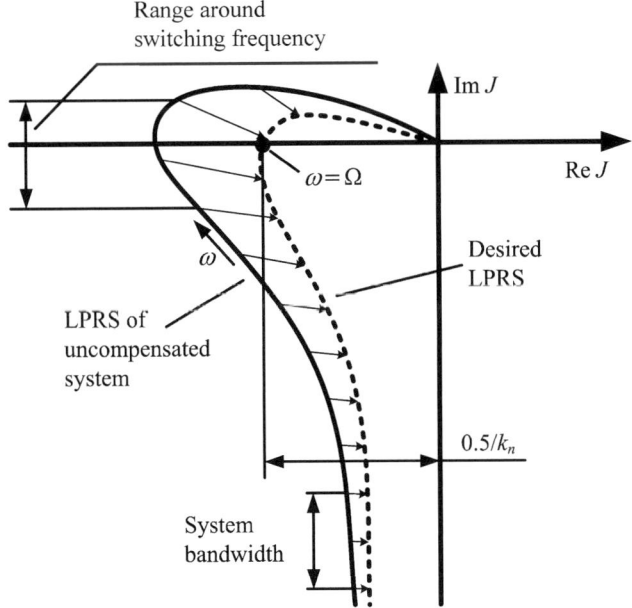

Fig. 6.4. Desired LPRS configuration

tem in the specified bandwidth do not change). The methodology of design in the system is as follows. First, an uncompensated servomechanism consisting of the plant and the relay is analyzed. At this stage, the frequency characteristics and the LPRS of the plant, the frequency of self-excited oscillations, and the frequency response of the closed-loop system are computed.

Next, the requirements to the desired LPRS configuration are calculated on the basis of the results of the uncompensated servomechanism analysis. Namely, a desired location of the point of intersection between the LPRS and the straight line "$-\pi b/(4c)$" and the corresponding frequency of self-excited oscillation is determined. This is done by applying formulas (6.3), (2.4) and the results of the analysis of the uncompensated servomechanism.

Depending on the application of the servomechanism, its performance can be specified in a few different ways. Consequently, there are various ways of generating the desired LPRS. Assume that the compensator is not supposed to change the frequency response of the open-loop system in the bandwidth. Then the desired LPRS location is determined by

$$\begin{cases} \mathrm{Re} J_d(\Omega_d) > -0.5/k_{nd} \\ \Omega_1 < \Omega_d < \Omega_2 \\ \mathrm{Im} J_d(\Omega_d) = -\pi b/(4c) \end{cases} \quad (6.7)$$

where "d" denotes "desired," k_{nd} is the desired value of gain k_n of the relay, Ω_1, Ω_2 are the specified range for the switching frequency, b is the specified value of the hysteresis (usually small value), and c is the output level of the relay (control).

Considering the idea of changing the LPRS shape (location at higher frequencies), we must assess what kinds of filters and signals are available. We might consider the use of low-pass, high-pass, phase-lead, phase-lag, lead-lag, lag-lead, band-pass, and band-rejecting filters and two possible points of connection: in the output signal (error signal), which provides the cascade connection of the filter (in series with the plant), and in the output of the relay, so that the output of the filter is summed with the system input, which is a parallel connection of the filter (parallel with the plant). Among possible connections and filters, two types of compensators perform the described function of the LPRS transformation: (1) the cascade connection with the use of the phase-lag filter, and (2) the parallel connection with the use of band-pass filter.

The Cascade Compensation. The output of the servomechanism is always available. Therefore, cascade compensation can be implemented in most cases. The circuit connected in series with the plant must force the LPRS to have a desired location. This can be done using elements with the transfer function

$$W(s) = \frac{T_1 s + 1}{T_2 s + 1},$$

6.3 Compensator design in the relay electro-pneumatic servomechanism

where
$$\omega_{max} < 1/T_2 < 1/T_1 < \Omega,$$

ω_{max} is the upper boundary of the specified input signal bandwidth.

This compensator does not strongly affect the frequency response of the open-loop system at frequencies within the bandwidth. It changes the location of the LPRS at higher frequencies only (beginning at a certain frequency, the LPRS of the compensated system can be approximately calculated as the product of the plant LPRS and the coefficient $T_1/T_2 < 1$).

Of course, other dynamic filters of higher order with similar amplitude-frequency response can be used for such compensation. One of the features of such cascade compensation is that it causes the frequency of self-excited oscillations to *decrease*. An example of the cascade compensation is considered below.

The Parallel Compensation. The output of the relay is always available and can be used for parallel compensation.

The circuit connected in parallel with the plant must force the LPRS to have a desired location (Fig. 6.4). The transfer function of the open-loop system is calculated as a sum of the transfer functions of the plant and the compensator. As a result, in the case of parallel compensation, the LPRS of the open-loop system is calculated as a sum of the plant LPRS and the compensator LPRS (see Theorem 2.1). Yet the addition of the compensator must not change the frequency response of the open-loop system at frequencies within the bandwidth.

Such properties are typical of the band-pass filters with the bandwidth encompassing the frequency of self-excited oscillations of the system. Formulas of the LPRS of some band-pass filters are presented in Table 2.1. The LPRS of second-order band-pass filters are depicted in Fig. 6.5. One of the features

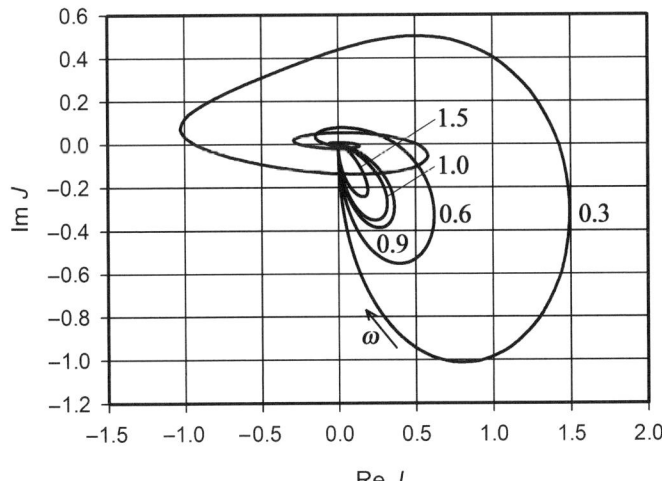

Fig. 6.5. The LPRS $J_n(\omega)$ of band-pass filters $W(s) = s/(s^2 + 2\xi s + 1)$ for $\xi = 0.3 - 1.5$

of parallel compensation is that it can cause the frequency of self-excited oscillations to *increase*.

It is worth mentioning that Table 2.1 and Fig. 6.5 show the normalized LPRS $J_n(\omega)$ (computed for unity gain and time constants). To obtain the LPRS for the transfer function

$$W(s) = \frac{KTs}{T^2 s^2 + 2\xi T s + 1}$$

(including the case of the general form with $\xi \geq 1$), we have to recalculate the LPRS according to

$$J(\omega) = K\, J_n(T\omega),$$

where $J_n(\omega)$ is the normalized LPRS of the band-pass compensator.

An example of parallel compensation is considered below.

6.4 Examples of compensator design in the relay electro-pneumatic servomechanism

Consider cascade and parallel compensation of the electro-pneumatic servomechanism (Fig. 6.6).

In Fig. 6.6, future cascade and parallel compensators are displayed as dashed lines.

Using formulas (6.1), (6.2), we calculate the LPRS of the uncompensated servomechanism. The model parameters are as follows: $K_1 = 160$, $K_2 = 105$, $K_3 = 0.4$, $T_1 = 0.002\,\text{s}$, $T_2 = 0.003\,\text{s}$, $T_3 = 0.04\,\text{s}$. The LPRS plot is depicted in Fig. 6.7 (plot #1). We draw the horizontal line "$-\pi b/(4c)$" ($c = 1$, $b = 0.01$). We now solve equation (6.3) to calculate the frequency Ω_0 of self-excited oscillations in the uncompensated servomechanism. It yields $\Omega_0 = 429.4$ rad/s.

Let the specified bandwidth of the servomechanism be [1Hz; 8Hz] (or [6.28 rad/s; 50.2 rad/s]) with the maximum amplitude of the harmonic input $A_{\max} = 1\,\text{V}$. According to (2.4), we calculate the equivalent gain of the relay: $k_n = 0.768$. The amplitude and phase frequency response of the closed-loop uncompensated servomechanism is depicted in Fig. 6.8. It is calculated on the basis of the linearized model (solid line) and as the Fourier series analysis of the system output $y(t)$ obtained through computer simulation (dots). Very good agreement is observed within the system bandwidth.

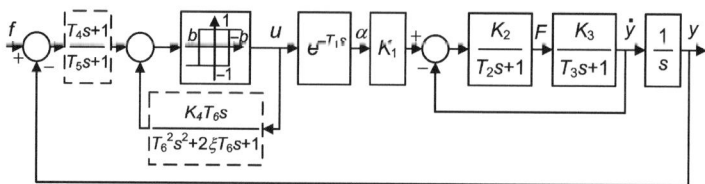

Fig. 6.6. Cascade and parallel compensation of relay servomechanism

6.4 Examples of compensator design   133

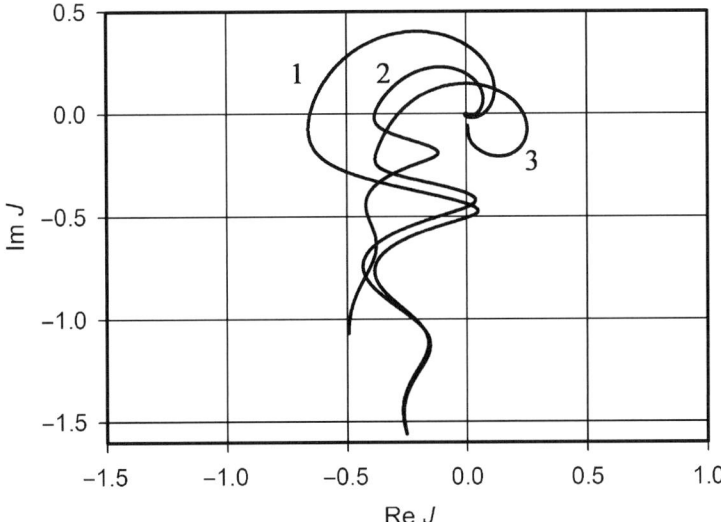

Fig. 6.7. The LPRS of electro-pneumatic servomechanism (1, uncompensated; 2, cascade-compensated; 3, parallel-compensated)

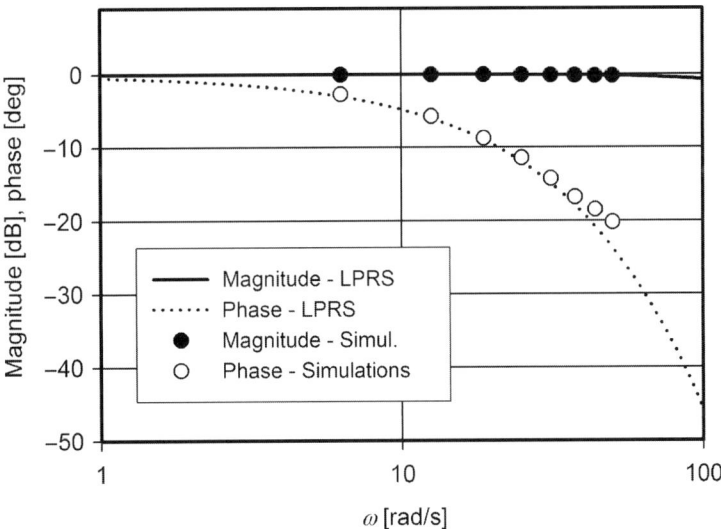

Fig. 6.8. Closed-loop frequency response of the uncompensated servomechanism

We specify the requirements to the desired LPRS location as follows, which means that the desired value of the open-loop gain is twice the gain of the uncompensated system:

$$\mathrm{Re}J(\Omega) = 0.5 \ \mathrm{Re}J_d(\Omega_0), \tag{6.8}$$

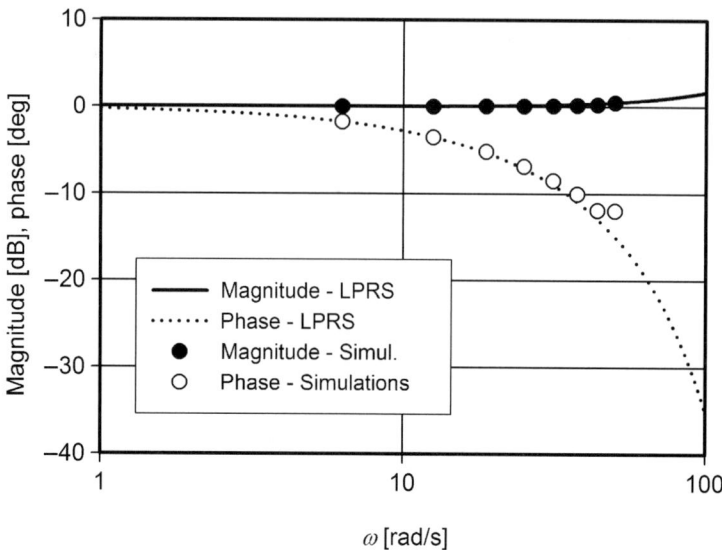

Fig. 6.9. Closed-loop frequency response of the cascade-compensated servomechanism

$$0.85\,\Omega_0 < \Omega < 1.15\,\Omega_0.$$

The last inequality limits the change in the frequency of self-excited oscillations due to the compensation.

The Cascade Compensation. Formula (6.8) leads to $T_4/T_5 < 0.5$. We choose $T_4/T_5 = 0.5$ and place the range $[1/T_5; 1/T_4]$ in the middle of the range $[50.2\,\text{rad/s}; 429.4\,\text{rad/s}]$. Thus, $T_4 = 0.0043\,\text{s}$, $T_5 = 0.0086\,\text{s}$. The LPRS of the compensated servomechanism and the frequency response of the closed-loop compensated servomechanism are depicted in Fig. 6.7 (plot #2) and Fig. 6.9 respectively (solid line, linearized model; dots, the result of process simulation and Fourier analysis). The performance enhancement is achieved and revealed as $-12.0°$ of maximum phase lag for the compensated system within the bandwidth, versus $-20.2°$ of phase lag for the uncompensated system.

The Parallel Compensation. We choose $\xi = 0.6$ and $T_6\Omega = 1.2$, which corresponds to the point of the normalized LPRS $J_n(T_6\Omega) = 0.602 - j0.423$ ($\arg J_n(T_6\Omega) = -35.1°$, i.e., the angle at which the magnitude of J_n is maximized).

The specified point of intersection of the desired LPRS and the straight line "$-\pi b/(4c)$" is projected to the LPRS of the plant at the angle of $180° - 35.1° = 144.9°$, and the frequency Ω is obtained as a corresponding point of the plant LPRS:

$$\Omega = 475.5\,\text{rad/s}.$$

T_6 is then calculated as $T_6 = 1.2/\,\Omega = 0.0025\text{s}$.

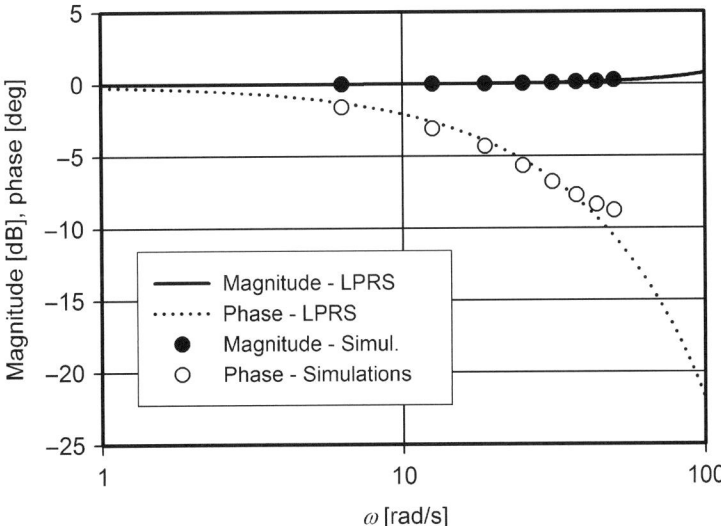

Fig. 6.10. Closed-loop frequency response of the parallel-compensated servomechanism

The LPRS of the compensated system is depicted in Fig. 6.7 (plot #3). The frequency response of the compensated system is depicted in Fig. 6.10 (solid line, linearized model; dots, the result of process simulation and Fourier analysis). We note that performance enhancement is achieved and revealed as $-8.8°$ of maximum phase lag for the compensated system within the bandwidth versus $-20.2°$ of phase lag for the uncompensated system.

The design of the compensator carried out above is a first iteration of the design procedure, and the performance of the system can be further improved by applying an optimization procedure to the filter parameters. But even this first step brings a significant enhancement to the system performance.

6.5 Compensator design in the relay electro-pneumatic servomechanism with the use of the LPRS of a nonlinear plant

The use of the nonlinear model of the servomechanism offers higher accuracy of design. The most significant precision enhancement can be achieved by using the nonlinear model of the solenoid valve and considering the Coulomb friction of the piston motion inside the cylinder. The use of the time delay model of the solenoid valve does not allow for the impact of the solenoid travel time. The Coulomb friction between the cylinder and the piston can also be significant and, therefore, noticeably affect the system dynamics.

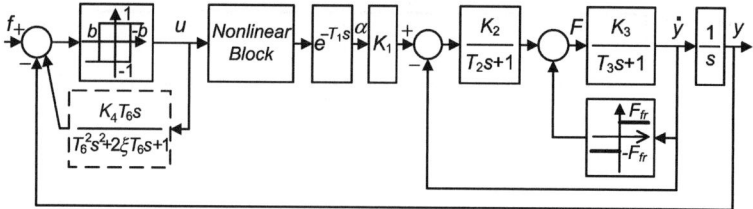

Fig. 6.11. Nonlinear servomechanism with parallel compensation

The design of the servomechanism with the use of the linear model of the plant was considered in Chapter 5. Now we consider a nonlinear model of the uncompensated servomechanism (Fig. 6.11).

The solenoid model is represented by a nonlinear block plus a time delay. This model reflects the finite and constant travel time of the air valve. The model is given as follows:

$$\alpha(t) = \begin{cases} -\alpha_{\max}, & \text{if } u = c \text{ and } t^+ \leq t < t^+ + \tau_1 \\ \frac{2\alpha_{\max}}{\tau_2}(t - t^+ - \tau_1) - \alpha_{\max}, & \text{if } u = c \\ \quad \text{and } t^+ + \tau_1 \leq t \leq t^+ + \tau_1 + \tau_2 \\ \alpha_{\max}, & \text{if } u = c \text{ and otherwise} \\ \alpha_{\max}, & \text{if } u = -c \text{ and } t^- \leq t < t^- + \tau_1 \\ -\frac{2\alpha_{\max}}{\tau_2}(t - t^- - \tau_1) + \alpha_{\max}, & \text{if } u = -c \\ \quad \text{and } t^- + \tau_1 \leq t \leq t^- + \tau_1 + \tau_2 \\ -\alpha_{\max}, & \text{if } u = -c \text{ and otherwise,} \end{cases} \quad (6.9)$$

where t^+, t^- is the time of the last switch of the relay from "$-$" to "$+$" and from "$+$" to "$-$" respectively, τ_1 is the time delay of the solenoid, and τ_2 is the air valve travel time from position $-\alpha_{\max}$ to position $\alpha_{\max}$.

The above model of the solenoid reflects the fact that the shape of the armature motion signal in the relay pneumatic servo system is close to trapezoidal. Consequently, the above nonlinear block transforms the square pulse signal into the trapezoidal signal.

Suppose the system parameters have the following values: $c = 1$, $b = 0.1$, $K_1 = 160$, $K_2 = 105$, $K_3 = 0.41$, $\tau_1 = 0.0004$ s, $\tau_2 = 0.0032$ s, $\alpha_{\max} = 1$, $T_1 = 0.002$ s, $T_2 = 0.003$ s, $T_3 = 0.041$ s, $F_{fr} = 20$ N.

The system in Fig. 6.11 can be viewed as one with an integrating plant like that in Fig. 2.2. Assume that a periodic process exists in the system. For that reason, no additional elements connected in parallel are needed to compute the LPRS of this nonlinear plant (this approach was described in Section 2.9). A simulation of this system with a small (f_0=5% of its maximal value) constant input and hysteresis b varied within range $b \in [-1.9; 0.5]$ was carried out (aimed at generating self-excited oscillations of various frequencies), and values of $\sigma(t)$ in a steady periodic motion were recorded in a file. At

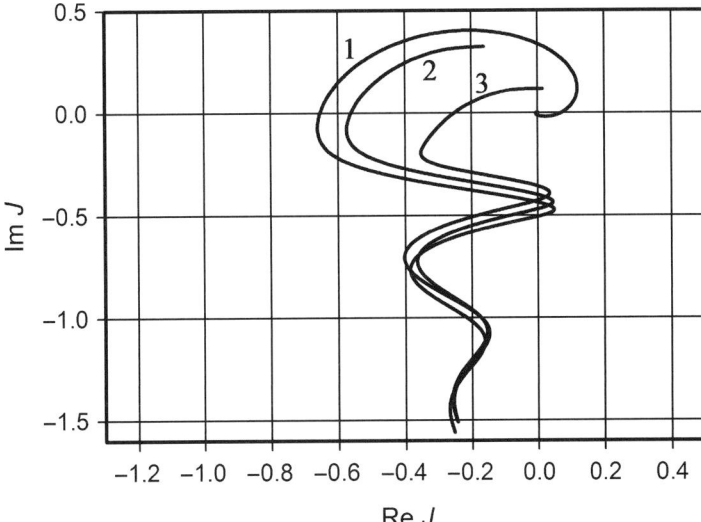

Fig. 6.12. The LPRS of the servomechanism (1, uncompensated with linear plant; 2, uncompensated with nonlinear plant; 3, parallel-compensated)

each frequency of oscillations, the LPRS was calculated as per the following formula:

$$J(\Omega) \approx -0.5\frac{\sigma_0}{u_0} + j\frac{\pi}{4c}(f_0 - \sigma(t))|_{t=0}. \tag{6.10}$$

The result of this simulation and calculations are presented in Fig. 6.12 (plot #2). The LPRS calculated on the basis of the linearized model of the plant is depicted in the same figure for comparison (plot #1). We see that the high-frequency segments of the two LPRS differ more than the low-frequency segments. This is a result of the air valve travel time, which has a stronger effect on the resulting LPRS at higher frequencies.

Compensation by a parallel band-pass filter, tuned so that its bandwidth encompasses the frequency of the oscillations, can be used for system performance enhancement in the same way as in the previous section for the linear plant. Applying the technique and formulas of the LPRS of a band-pass filter, we calculate the parameters of the compensating filter as follows:

$$K_4 = 0.378, \ T_6 = 0.0025s, \xi = 0.6.$$

We note that the nonlinear features of the plant allow for more accurate LPRS calculation and, as a result, more accurate compensator design. If the compensator were designed on the basis of the linearized model of the plant (plot #1 in Fig. 6.12), it might result in small stability margins for the system (the LPRS of the compensated system might approach the origin too near and even cross the line "$-\pi b/(4c)$" in the right half-plane). The LPRS of the compensated system is depicted in Fig. 6.12 as plot #3. The frequency

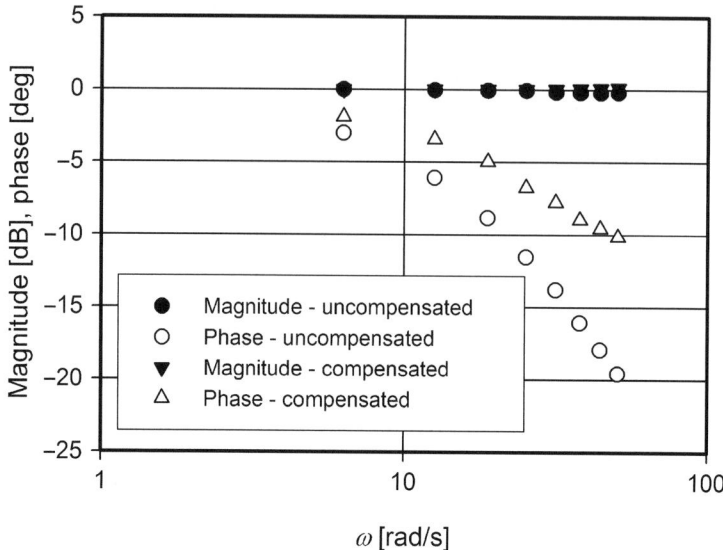

Fig. 6.13. Servomechanism closed-loop dynamics frequency response (nonlinear model of the plant; uncompensated and parallel-compensated)

responses of the uncompensated and compensated servomechanisms are depicted in Fig. 6.13. We note that the performance enhancement, estimated as a phase lag reduction, is significant.

6.6 Conclusions

We consider the application of the LPRS method to analysis of dynamics and design of compensators for relay electro-pneumatic servomechanisms, and we present a methodology of analysis and design. We demonstrate that the use of the LPRS method not only allows for a precise analysis of the complex dynamics of the relay electro-pneumatic servomechanism, but also provides a convenient tool for compensator selection and design. Beside the linear model of the servomechanism, the concept of the LPRS of a nonlinear plant is used for the compensator design in the relay electro-pneumatic servomechanism with the Coulomb friction and nonlinear model of the solenoid and air valve dynamics. Despite the increase of the amount of computations, the use of the LPRS leads to an efficient design of the system.

7
Relay feedback test identification and autotuning

7.1 The relay feedback test

Proportional-integral-derivative (PID) control is the main type of control used in the process industry. PID controllers are usually implemented as configurable software modules within distributed control systems (DCS). The DCS configuration software is constantly evolving and giving developers many new features. One of most useful features is the controller autotuning feature. This trend can be seen in the development of new releases of such popular DCS software as Honeywell Experion PKS and Emerson DeltaV. Despite the existence of a large number of tuning algorithms, there is still a need for simple and precise loop tuning algorithms that can be embedded as additional autotuning add-ons in the PID controllers of DCS. The requirements of the controller autotuners are simplicity, precision, and robustness.

One of the most convenient tests on the process is the relay feedback test proposed in [5]. This method has received a lot of attention from both industry and the worldwide research community. It has become a starting point for a number of directions of research as well. In comparison to the original approach by Ziegler and Nichols [109], which was aimed at obtaining the values of the *ultimate gain* and *ultimate frequency* (i.e., the minimal gain that brings the system to the state of self-excited oscillations and the frequency of those oscillations), it was proposed in [70] that the relay feedback test be used for process parameters identification. This idea was further developed and extended to various models and types of processes. The survey of available tuning methods and techniques based on the relay feedback test is presented in [7]. However, despite the obvious success of the relay feedback test in autotuning and identification, it can lead to significant errors. The errors come from the model of the oscillations based on the application of the approximate describing function method. There have been a few attempts to overcome this source of inaccuracy which have solved the problem to some degree. In [59], for example, it is proposed that the amplitude of the oscillations be used in addition to the imaginary part of the Tsypkin locus [94]. This results in a precise model for two simple transfer functions. In [60] and [61], it is shown how the parameters of first-order and second-order process transfer functions with time delay can be found exactly using the A-locus [9] from measurements of the asymmetric limit cycle. In [73], exact parameters of first-order and second-order plus dead-time models are obtained from measurements of the asymmetric limit cycle. In [74], a relay feedback and wavelet-based method for estimation of unknown processes is proposed. In [106], it is proposed that the saturation function should be used instead of the relay nonlinearity, which transforms the square wave into a nearly sinusoidal one and, consequently, brings the test to the limitations of the describing function method.

However, the problem has not been fully solved. The fundamental obstacle here is the absence of a simple and precise model of the oscillations in a relay feedback system, which would lead to simple calculations suitable for DCS applications. The LPRS method offers an opportunity to further solve the problem of accuracy. Although the LPRS provides an exact model only for an infinitesimally small asymmetry of the oscillations, experiments on a number of plants and simulations show that linear approximation of the propagation through the relay function is also precise within a certain non-small finite range. This ensures sufficiently high accuracy of the application of the LPRS method to autotune identification.

7.2 The LPRS and asymmetric relay feedback test

A closer look at the LPRS definition (2.2) shows that the LPRS is a characteristic that can be measured from the asymmetric relay feedback test. Indeed, the real part is defined as a ratio of two constant values (it is defined as a limit of the ratio, and the problem of the accuracy of this measurement is considered below), and the imaginary part is equal to the hysteresis value (with a coefficient). Hence, there are two factors available that are important for the solution of the identification problem. First, the LPRS can be computed from the process model, and second, it can be measured from the relay feedback test. Therefore, the identification methodology is based on the matching of the computed LPRS to the measured LPRS. The identification is carried out as follows. An asymmetric relay feedback test is run over the process, and the length of the positive control pulse θ_1, the negative control pulse θ_2 (see Fig. 1.5), and the average on the period process output y_0 are measured (the constant input f_0, hysteresis b, and the amplitude of the relay c are given parameters). On the basis of the measured three values, the following parameters are computed. The frequency of the oscillations is computed from the obvious formula

$$\Omega_m = \frac{2\pi}{\theta_1 + \theta_2},$$

and the average control signal is computed from the pulses length measurements as
$$u_0 = c\frac{\theta_1 - \theta_2}{\theta_1 + \theta_2}.$$

With those values available, the following two equations for two unknown process parameters and a formula for the process static gain K can be written (we assume no disturbance):

$$\operatorname{Re} J(\Omega_m) = -\frac{1}{2}\frac{f_0 - y_0}{u_0} \tag{7.1}$$

$$\operatorname{Im} J(\Omega_m) = -\frac{\pi b}{4c} \tag{7.2}$$

$$K = \frac{y_0}{u_0}. \tag{7.3}$$

Therefore, one relay feedback test provides three parameters of the process model — via the solution of (7.1), (7.2), (7.3), with the LPRS computed as per (2.12), (2.19), (2.32) or using other techniques. In the case of more than three unknown parameters, a few relay feedback tests with different values of the hysteresis b must be carried out. Each test provides one frequency point of the LPRS and, consequently, each additional test allows identification of two additional parameters.

This is the general idea of the LPRS-based identification. One particular case that involves a very common process model — the first-order plus dead-time dynamics — is considered below in detail.

7.3 Methodology of identification of the first-order plus dead-time process

Many industrial processes can be precisely approximated by the first-order plus dead-time (FODT) transfer function:

$$W(s) = \frac{Ke^{-\tau s}}{Ts + 1} \tag{7.4}$$

The LPRS formula for FODT dynamics was derived in Section 2.7 with the LPRS formula given by:

$$J(\omega) = \frac{K}{2}(1 - \alpha e^\gamma \operatorname{csch}\alpha) + j\frac{\pi}{4}K\left(\frac{2e^{-\alpha}e^\gamma}{1 + e^{-\alpha}} - 1\right) \tag{7.5}$$

where $\alpha = \frac{\pi}{T\omega}$ and $\gamma = \frac{\tau}{T}$. With the values θ_1, θ_2, and y_0 measured from the asymmetric relay feedback test, and given the values of the constant input f_0, relay amplitude c, and hysteresis b, we now formulate the identification problem as the solution of the following set of equations:

$$\operatorname{Re} J(\Omega_m) = \frac{K}{2}(1 - \alpha e^{\gamma} \operatorname{csch} \alpha) = -\frac{1}{2}\frac{f_0 - y_0}{u_0} \qquad (7.6)$$

$$\operatorname{Im} J(\Omega_m) = \frac{\pi}{4}K\left(\frac{2e^{-\alpha}e^{\gamma}}{1+e^{-\alpha}} - 1\right) = -\frac{\pi b}{4c} \qquad (7.7)$$

$$K = \frac{y_0}{u_0} \qquad (7.8)$$

where $\Omega_m = 2\pi/(\theta_1 + \theta_2)$, $u_0 = c(\theta_1 - \theta_2)/(\theta_1 + \theta_2)$, $\alpha = \frac{\pi}{T\Omega_m}$ and $\gamma = \frac{\tau}{T}$.

Because K can be calculated separately, according to (7.8), we have to solve the two equations (7.6) and (7.7) with two unknown values T and τ. By expressing e^{γ} from (7.6) and substituting in (7.7), equations (7.6)–(7.8) can be reduced to one equation with one unknown variable α as follows:

$$\frac{y_0}{f_0} = \frac{1 - e^{-\alpha}}{\alpha}. \qquad (7.9)$$

Consider the solution of equation (7.9) in more detail. We compute and plot the function $\psi(\alpha)$ (Fig. 7.1):

$$\psi(\alpha) = \frac{1 - e^{-\alpha}}{\alpha}$$

It is possible to find a suitable approximation for the inverse function $\alpha = \alpha(\psi)$. However, here we can tabulate this function within the range $\psi \in [0.2; 1]$ and use an interpolation for in-between values. The tabulation of $\alpha = \alpha(\psi)$ is given in Table 7.1. For the values ψ below 0.2, the following approximate formula can be used:

$$\alpha \approx 1/\psi. \qquad (7.10)$$

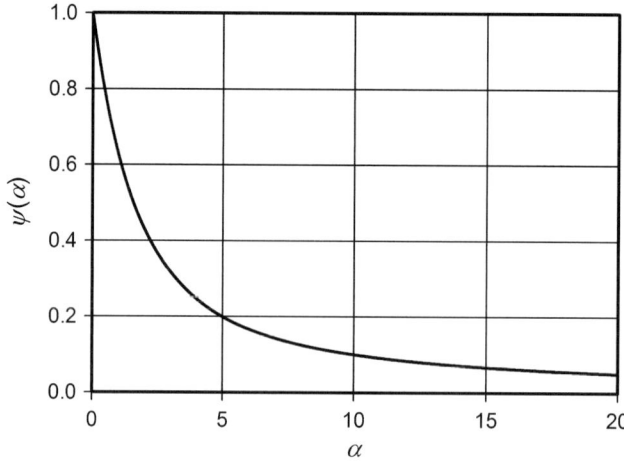

Fig. 7.1. Function $\psi(\alpha)$

Table 7.1. Tabulation of function $\alpha = \alpha(\psi)$

ψ	0.2	0.3	0.4	0.5	0.6	0.7	0.8	0.9	1.0
$\alpha = \alpha(\psi)$	4.965	3.197	2.232	1.594	1.126	0.761	0.464	0.215	0

Therefore, the identification is carried out using the following algorithm.

(a) The values of θ_1, θ_2, and y_0 are measured from the asymmetric relay feedback test and Ω_m and u_0 are calculated.

(b) The static gain K of the process is calculated as per (7.8).

(c) The equation (7.9) is solved for α by interpolating the data of Table 7.1 or using formula (7.10).

(d) Once the parameter α is found, the time constant T is calculated as

$$T = \frac{\pi}{\alpha \Omega_m}.$$

(e) Finally the dead time τ is calculated as

$$\tau = T \ln\left[\frac{1}{2}(e^\alpha + 1)\right].$$

The described algorithm is very easy to program and implement as an add-on to a PID controller. In spite of its simplicity, it can perform very well if the process is described by formula (7.4) adequately enough. It is, therefore, very suitable for an autotune identification. A few simulation examples are given below.

7.4 Analysis of potential sources of inaccuracy

Consider the following example that demonstrates the identification algorithm described above.

Example 7.1. Suppose the process is described by the first-order plus dead-time transfer function

$$W(s) = \frac{0.5e^{-0.5s}}{1.5s + 1}.$$

The parameters of the relay are chosen as follows: $c = 1$, $b = 0$. The constant input signal value is $f_0 = 0.1$. The relay feedback test produces the following parameters of the oscillations: positive and negative pulse duration $\theta_1 = 1.165$ s, $\theta_2 = 0.625$ s, and the average value of the process output $y_0 = 0.0754$. The process transfer function identified in formulas (7.6)–(7.8) is

$$W(s) = \frac{0.4998e^{-0.5130s}}{1.5059s + 1}.$$

The highest identification error is of the dead time (2.6%). The other two errors are much smaller: 0.04% for the gain and 0.39% for the time constant. All the error values are acceptable for the autotuning purpose, as the main source of error in the autotuners is the distinction between the underlying process model and the actual process dynamics. It should also be mentioned that the pulse duration in the relay feedback test is 65% for the positive pulse and 35% for the negative pulse. Therefore, the asymmetry of the control is significant. This substantiates the use of the real part of the LPRS (being defined as a limit) at non-small values of the input signal. This phenomenon is considered in more detail below.

In spite of the demonstrated accuracy, this example, of course, is not a proof of the efficiency of this algorithm, in particular, and of the whole methodology of matching the analytical and experimental LPRS, because the actual system response is always the result of a number of different factors. For that reason, potential sources of inaccuracy are considered below.

Besides the approximation of the function $\alpha = \alpha(\psi)$, there are two potential sources of errors within the considered methodology. The first one is the use of finite values for the $\mathrm{Re}J(\Omega_m)$ estimation while the LPRS is defined as a limit at $f_0 \to 0$. Another potential source of error is the measurement of the frequency of the oscillations Ω_m in the asymmetric test when the LPRS is defined only for the case of infinitesimally small asymmetry of the oscillations.

If we consider the value of u_0 as a function of σ_0, which is a result of the variation of f_0 within a certain range, we obtain the bias function. The derivative of this function at $\sigma_0 = 0$ is the equivalent gain of the relay. If, therefore, the bias function is linear within a certain range around the point $\sigma_0 = 0$, the equivalent gain is a good approximation and the calculation of $\mathrm{Re}J(\Omega)$ using finite values results in a small error. Simulations prove that the bias functions for many plants are virtually linear within the range of up to 80% of u_0. For the given process approximation, the bias functions for $\gamma = \tau/T = 0.2$ (plot #1) and $\gamma = \tau/T = 0.5$ (plot #2) are depicted in Fig. 7.2. One can see that in both cases, the bias functions are virtually linear within $[-80\%;+80\%]$ and $[-70\%;+70\%]$ respectively, of the input f_0 maximum span $f_{0MAX} = cK$, which is the maximal value of f_0 that still allows for the existence of oscillations in the system. Note that the bias of the functions is symmetric about the origin. The actual span of f_0 used for obtaining the plots is 0–90% of $f_{0MAX} = cK$.

Another contributing factor in the error is the measured frequency of the oscillations Ω_m, which is not equal to the frequency of the symmetric oscillations. Again, simulations prove that within the range of up to 40% of u_0, the frequency does not significantly change. An indirect substantiation of this observation is the fact that the derivative of the frequency of the oscillations with respect to the input f_0 in the point $f_0 = 0$ is zero (see Appendix). Additionally, there always exists the option of setting the input to zero and

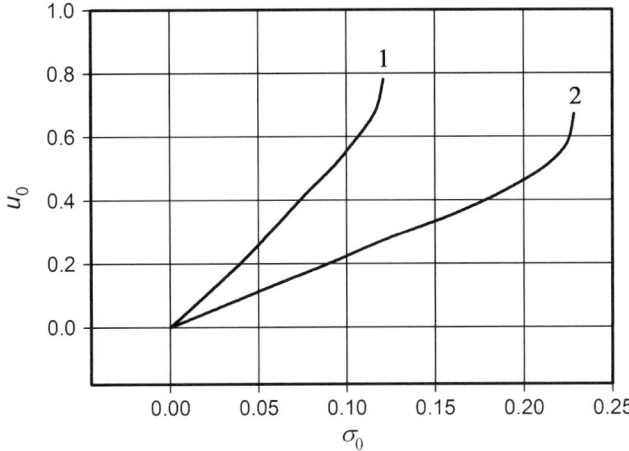

Fig. 7.2. Bias functions for relay system with first-order plus dead-time plant

measuring the frequency of the symmetric oscillations. Therefore, there exists a certain optimal range of f_0 settings that provides a sufficiently precise result. From simulations and experiments, the optimal range is the one that results in the average control u_0 10%–30% of its maximal value ($u_{0\,\max} = c$). This is sufficiently large to obtain precise measurements of $\mathrm{Re}\,J(\Omega)$ and sufficiently small to obtain a good estimate of frequency of symmetric oscillations.

7.5 Performance analysis of the identification algorithm

We consider a few different situations that commonly occur in the operation of industrial autotuners and assess performance of the described identification algorithm via simulations.

Example 7.2. Suppose the process is described by the same first-order plus dead-time transfer function as the example above:

$$W(s) = \frac{0.5e^{-0.5s}}{1.5s + 1}.$$

Ideal conditions. We randomly generate two parameters uniformly distributed within the ranges $T \in [0.1; 5.1]$, $\tau \in [0.1; 5.1]$, which cover the range $\tau/T \in [0.02; 51]$ (gain K is considered constant as it can be identified exactly per (7.8)); we run the identification algorithm; and we assess the accuracy of identification through standard (root mean square) deviations. The accuracy of identification in 1000 tests with random T and τ values are as follows: the standard deviations are $\sigma_T = 2.76\%$ and $\sigma_\tau = 0.82\%$. The existence of

Table 7.2. Identification accuracy (standard deviation) in noisy conditions

	$a_{N\,MAX} = 5\,\%$	$a_{N\,MAX} = 10\,\%$	$a_{N\,MAX} = 15\,\%$	$a_{N\,MAX} = 20\,\%$
σ_T [%]	3.32	4.00	4.47	6.08
σ_τ [%]	1.08	1.17	1.55	2.05

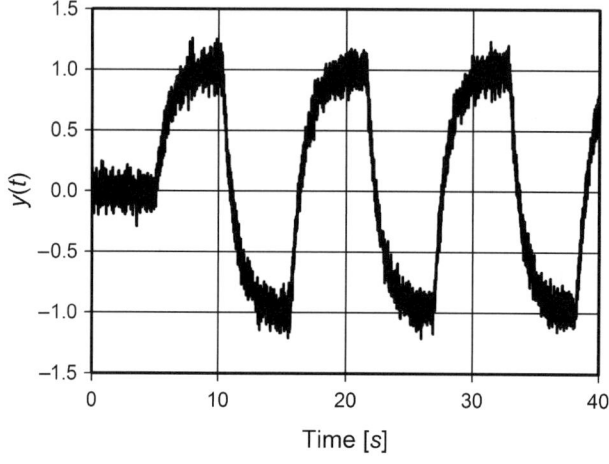

Fig. 7.3. Relay feedback test over process $e^{-5s}/(s+1)$ with noise component $a_{NMAX} = 25\%$

the identification errors is a result of the use of the equivalent gain as an approximation for the bias functions in Fig. 7.2 and the dependence of the frequency on f_0.

Noisy measurements. Let us inject the measurement noise of different amplitudes and assess the accuracy of identification. In Table 7.2, identification errors for T and τ are presented at different levels of white noise injected at the system output $y(t)$. The statistics of 100 tests are given for each case. Noise is characterized by its maximal amplitude measured with respect to the amplitude of $y(t)$. An example of this test over the process $e^{-5s}/(s+1)$ with noise component $a_{NMAX} = 25\%$ is presented in Fig. 7.3. The transfer function identified from the test is $e^{-4.912s}/(1.072s + 1)$.

External constant load (disturbance). In the practice of controller tuning in process industries, the relay feedback test over the process can only be implemented in an incremental way. This involves initially bringing the process controlled by a PID controller (not yet tuned or at least not optimally tuned) to a steady state. After that, the relay feedback test is applied to the process via controller output changes from the steady state by $\pm c$. However, if an external disturbance is applied to the process and this disturbance is constant, its effect is the same for the initial (prior to the test) control and during the test. Therefore, the disturbance is compensated for by the correct initialization before the test. The correctness of the initialization involves

Table 7.3. FODT approximation of higher-order dynamics

Test #	Actual process	FODT model	FODT model as per [61]
a	$\exp^{-2s}\dfrac{1}{(2s+1)^2}$	$\exp^{-3.112s}\dfrac{1}{4.150s+1}$	$\exp^{-2.998s}\dfrac{1}{4.271s+1}$
b	$\exp^{-2s}\dfrac{1}{(2s+1)^5}$	$\exp^{-7.675s}\dfrac{1}{6.641s+1}$	$\exp^{-7.420s}\dfrac{1}{7.066s+1}$
c	$\exp^{-0.5s}\dfrac{1}{(s+1)(s^2+s+1)}$	$\exp^{-3.363s}\dfrac{1}{1.301s+1}$	$\exp^{-2.112s}\dfrac{1}{1.296s+1}$
d	$\exp^{-s}\dfrac{-s+1}{(s+1)^5}$	$\exp^{-5.838s}\dfrac{1}{2.782s+1}$	$\exp^{-5.082s}\dfrac{1}{2.292s+1}$

bringing the process to a steady state as precisely as possible before starting the test. In practice, disturbance may vary. Therefore, it is important that the test be quick enough, so that the changes of the disturbance during the test are small, since it was assumed above that the value of external load would not change during the test.

Higher-order process model. Also, we analyze robustness of identification with respect to the existence of higher-order dynamics of the process. We run the set of tests over the same higher-order dynamics as in [61], in which the actual higher-order dynamics are identified as a FODT model. The results are presented in Table 7.3. Results of [61] are also given for reference. Nyquist plots for each of those transfer functions are given in Figs. 7.4 and 7.5. One can see that there is high accuracy of approximation when the phase characteristic is close to $-180°$ (except for case "c" for which the process is oscillatory, in which case, the FODT is not a good approximation).

7.6 Tuning algorithm

Tuning criterion. There are a number of tuning criteria that are used for selecting optimal settings of PID controllers. Among the most well-known are minimum of integral absolute error (IAE) and integral time absolute error (ITAE). It is worth noting that those criteria are time-domain criteria and represent certain characteristics of the step response of the closed-loop system. On the other hand, the most easily measurable and most important characteristic of the step response is the value of overshoot. In other words, neither the above-mentioned criteria account for the overshoot value directly, nor is the overshoot included in these criteria as a constraint. In practical applications, even if the tuning is optimal in accordance with a certain criterion but the overshoot exceeds a desirable value, the choice is always in favor of the tuning rules that account for the overshoot constraint.

However, the overshoot cannot serve as a criterion of optimization, since it is related to stability rather than to performance. In this section, the overshoot is considered a constraint, which implies the use of a certain criterion

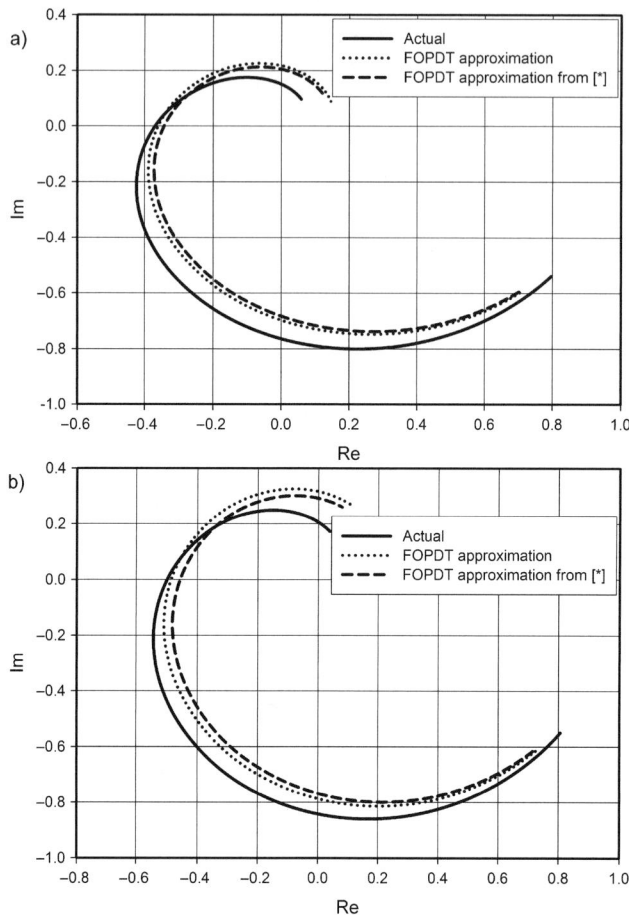

Fig. 7.4. Nyquist plots for options (a), (b) as per Table 7.3; [61] is denoted as [*]

that does not involve overshoot. Consider this criterion and assume for simplicity that the controller is only a proportional gain. The problem of tuning in this case is the problem of finding the maximum value of the gain that satisfies the overshoot constraint, as the higher values of the gain provide better disturbance rejection properties. In other words, there is a trade-off between the desired overshoot and the attainment of the maximum value of the controller gain. We note that this trade-off is resolved in a very simple way in this situation: we reach the maximum of the gain subject to the constraint. Suppose now that the controller is a PI-controller. Our objective is to increase both the proportional gain and the integral gain and to satisfy the constraint for the overshoot. However, the fact that we have to manipulate two gains does not allow for a unique solution. We need to reformulate the original objective of

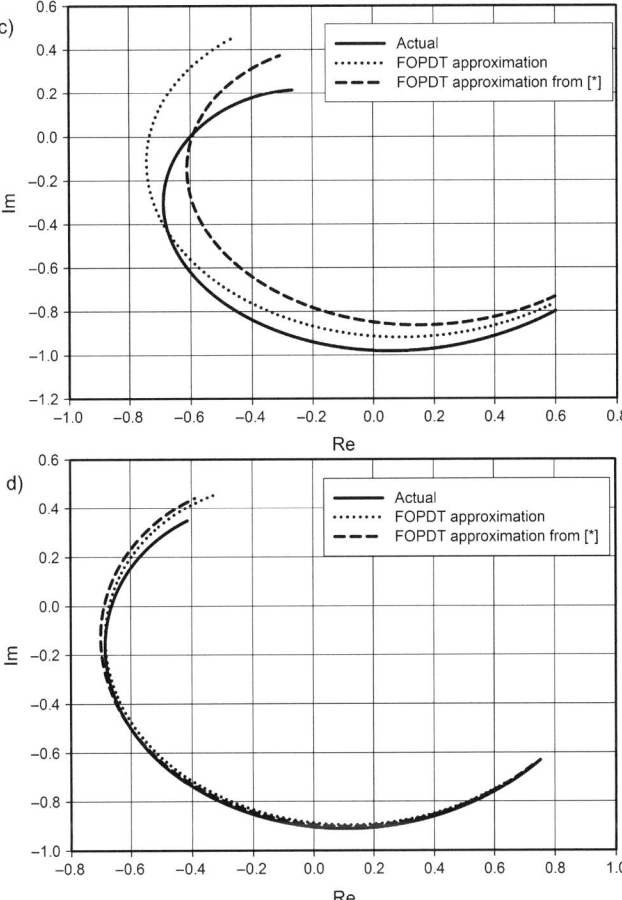

Fig. 7.5. Nyquist plots for options (c), (d) as per Table 7.3; [61] is denoted as [*]

maximizing the proportional gain into a different criterion, which also works for two and three gains. This is the criterion of minimal settling time. We note that in the case of the proportional gain, the maximum value of the gain also provides the minimum settling time. However, the former cannot be applied to the case of several parameters, whereas the latter can.

Therefore, we formulate and use the tuning criterion as the minimal settling time subject to satisfying the constraint on the overshoot value,

$$t_{set}(K_p, K_i, K_d) \to \min \tag{7.11}$$

$$\max_{t \in [o;\infty]} y(t) = 1 + y_{os}/100 \tag{7.12}$$

where $t_{set} = \begin{cases} t_{set} := y(t_{set}) = 1 - \Delta \quad \text{or} \quad y(t_{set}) = 1 + \Delta \\ y(t > t_{set}) \in [1 - \Delta; 1 + \Delta] \end{cases}$ is the settling time, Δ is the step response envelope width, and y_{os} is the overshoot value in %. In formulas (7.11) and (7.12), it is assumed a unity feedback and the unity step value.

PI-controller settings. The solution of this optimization problem for the underlying process transfer function $W(s) = \frac{Ke^{-\tau s}}{Ts+1}$ and the PI-controller leads to the following optimal settings of the controller.

For overshoot 5%, 10%, and 20%, the integrator normalized time constant can be computed as follows (respectively):

$$T_{0i} = 1.60\tau/T \qquad (7.13)$$

$$T_{0i} = 1.80\tau/T \qquad (7.14)$$

$$T_{0i} = 1.95\tau/T. \qquad (7.15)$$

The normalized values of the proportional gain are tabulated and presented in Table 7.4. (in-between values are found via interpolation).

Table 7.4 and formulas (7.13)–(7.15) provide the values of settings that apply to the transfer function with unity gain and time constant. For arbitrary parameter values, recalculation can be done using the following formulas, which give the scaled optimal solution that accounts for non-unity process gains and time constants. The proportional gain is

$$K_p = K_{0p}/K \qquad (7.16)$$

and the integrator gain is

Table 7.4. Normalized proportional gain settings K_{0p}

	$y_{os} = 5\%$	$y_{os} = 10\%$	$y_{os} = 20\%$
$\tau/T = 0.1$	5.203	5.957	7.177
$\tau/T = 0.2$	2.624	3.058	3.702
$\tau/T = 0.3$	2.823	2.120	2.564
$\tau/T = 0.4$	1.483	1.673	2.007
$\tau/T = 0.5$	1.294	1.419	1.683
$\tau/T = 0.6$	1.170	1.258	1.473
$\tau/T = 0.7$	1.082	1.148	1.329
$\tau/T = 0.8$	1.014	1.068	1.225
$\tau/T = 0.9$	0.964	1.008	1.146
$\tau/T = 1.0$	0.924	0.963	1.086
$\tau/T = 1.5$	0.808	0.833	0.915

$$K_i = \frac{1}{T_{oi}TK}. \qquad (7.17)$$

These settings of the PI-controller provide the fastest possible step response of the closed-loop system subject to the overshoot constraints.

Tuning algorithm. Therefore, the complete autotuning algorithm comprises the following steps.

(a) The values of θ_1, θ_2, and y_0 are measured from the asymmetric relay feedback test and Ω_m and u_0 are calculated.

(b) The static gain K is calculated as $K = y_0/u_0$.

(c) Equation (7.9) is solved for α with the use of the data of Table 7.1 or formula (7.10).

(d) Once the parameter α is found, the time constant T is calculated as $T = \pi/(\alpha\Omega_m)$.

(e) The dead time τ is calculated as $\tau = T\ln(0.5(e^\alpha + 1))$.

(f) The normalized proportional gain and integrator time constant are computed using Table 7.4 and formulas (7.13)–(7.15).

(g) The PI-settings of the controller are calculated using (7.16) and (7.17).

Consider the following example.

Example 7.3. Suppose the process is described by the following transfer function, which is considered unknown to the autotuner:

$$W(s) = \frac{0.5e^{-0.6s}}{0.8s^2 + 2.4s + 1}.$$

The parameters of the relay are chosen as follows: $c = 1$, $b = 0$. The constant input signal value is $f_0 = 0.1$. The objective is to design a PI controller for this process with the use of the first-order plus dead-time transfer function as an approximation of the process dynamics. After that, the PI controller must be tuned in such a way that the system produces the shortest possible settling time and the overshoot $\leq 10\%$ at the step response.

The following values of the oscillatory process are measured: the frequency of the oscillations $\Omega_m = 1.903$, the average value of the process output $y_0 = 0.0734$, and the average value of the control signal $u_0 = 0.1455$.

As per the described algorithm, the process parameters are identified as follows: $K = 0.5050$, $T = 2.5285$s, $\tau = 0.9573$s. The PI settings that provide the characteristics of the closed-loop system are $K_i = 1.349$ and $K_p = 3.503$. However, the actual system step response produces 12.5% overshoot. This example demonstrates the accuracy issues and trade-offs in the design of industrial autotuners.

7.7 Conclusions

Modern DCS do not usually provide a convenient environment for iterative computing (i.e., solution of algebraic equations). This is why representing identification algorithms by means of simple calculations is extremely important. A methodology of process identification based on the asymmetric relay feedback test and the LPRS method is considered in this chapter. The methodology involves fitting of the LPRS obtained analytically through the given model of the process to the points of the LPRS measured from the asymmetric relay feedback test. The methodology is presented in detail for the process approximation (underlying process model) being the first-order plus dead-time transfer function. We give simple analytical formulas and an algorithm of identification suitable for the autotuning for this process approximation. We consider examples and analyze the potential sources of identification errors. The described identification-tuning algorithm is patented [21] and implemented on Honeywell Experion PKS and TPS DCS platforms.

8

Performance analysis of the sliding mode–based analog differentiator and dynamical compensator

8.1 Transfer function "inversion" via sliding mode

The SM principle can be used to derive the input signal of a certain relatively low-order dynamical system from its output. This operation is equivalent to applying the reciprocal of the transfer function of this dynamical system to the output, which can be viewed as the transfer function "inversion." This is closely related to the problems of state observation and signal differentiation. In that respect, we analyze two devices: the SM differentiator and the SM compensator of sensor dynamics.

Obtaining the derivative of a signal is done with devices called differentiators which can be implemented either in hardware or software. Signal differentiation is a very important practical problem, since using a differentiator is normally a much cheaper option than using a sensor or a transmitter for obtaining variables such as velocity and rate of change. However, practical realizations of various differentiators reveal some fundamental limitations and drawbacks, such as the presence of time delay (or lag) and sensitivity to noise. Most of the efforts by designers of differentiators are aimed at mitigating these drawbacks.

There are a variety of devices and algorithms that can be used as differentiators. Beside the standard first or second finite difference algorithm, these include high-pass linear filters, Kalman filters, Luenberger observers, etc. The sliding mode principle can be used for the purpose of obtaining a derivative, too. The first-order SM differentiator is introduced in [97]. Its dynamics are described by the following equations:

$$\dot{y} = u \qquad (8.1)$$

$$u = c\ \text{sign}(f - y) \qquad (8.2)$$

where f is the input signal to be differentiated, and u is the output of the differentiator, which also contains a high-frequency component that needs to be filtered out. The integrator (8.1) can be considered a plant in the relay feedback system (8.1)–(8.2). These equations describe the first-order dynamics, and *ideal* SM occurs in the system. Because ideal SM occurs, the sliding variable $\sigma = f - y$ is zero and, consequently, $y = f$. As a result, $\dot{y} = \dot{f}$, and $u = \dot{f}$ (which, of course, is true only with respect to the *equivalent* control), i.e., the dynamics given by (8.1), (8.2) can be used for taking a derivative (computing the rate of change) of a signal. The derivative can be extracted from signal $u(t)$, which, however, in accordance with (8.2), can possess only two values: c and $-c$. Therefore, information about the derivative is modulated by the high-frequency component, and the latter needs to be filtered out through low-pass filtering.

8.2 Analysis of SM differentiator dynamics

Obviously, any practical implementation of the differentiator (8.1), (8.2) results in finite-frequency oscillations (chattering). This happens due to the existence of parasitic dynamics along with the principal dynamics. The character of parasitic dynamics depends on the differentiator realization. If, for example, the differentiator is implemented on operational amplifiers, the parasitic dynamics are associated with the distinction of the amplifier dynamics from a gain, which is revealed in the limited bandwidth of the operational amplifier. Another imperfection is a delay in the switching of the relay (8.2), which introduces a time delay in the differentiator loop. Both these imperfections result in the distinction of differentiator dynamics from the ideal differentiator, which is the subject of analysis in this chapter. A different situation occurs if a differentiator is implemented as a computing algorithm. In that case, the parasitic dynamics are associated with the discrete-time realization of the algorithm and the delays introduced into the differentiator loop. This type of imperfection is considered in what follows and in further detail in Chapter 9.

Let the differentiator (8.1), (8.2) be implemented on operational amplifiers as depicted in Fig. 8.1. In Fig. 8.1, if we assume $R_1 = R_2 = R_3$, the amplitude of the relay is determined by the limiting voltage of the Zener diodes, and the plant transfer function is

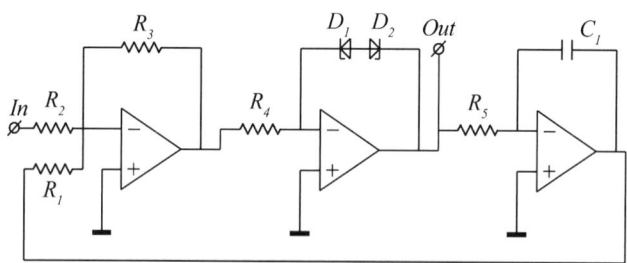

Fig. 8.1. Analog differentiator

8.2 Analysis of SM differentiator dynamics

$$W_p(s) = \frac{1}{R_5 C_1} \frac{1}{s}. \qquad (8.3)$$

However, as we noted above, the representation of the operational amplifier (with a feedback) as a gain is an idealization. This is satisfactory for most applications, but a more detailed description is needed for analysis of effects such as chattering in the system in Fig. 8.1. A more detailed description of the operational amplifiers would involve some dynamical properties of the operational amplifiers such as bandwidth and maximum rate of change of the output voltage. Let us assume that the parasitic dynamics for the first and third amplifiers are given as first-order dynamics, and the parasitic dynamics of the second amplifier (that has an on-off mode of operation) is given by the time delay τ. As a result, we can write the dynamics of the linear part of the relay servo system as follows:

$$W_l(s) = \frac{K e^{-\tau s}}{s(Ts+1)^2} \qquad (8.4)$$

where $K = \frac{1}{R_s C_1}$, $T = \frac{1}{2\pi f_{\max}}$, $f_{\max}$ is the amplifier bandwidth, $\tau = \frac{U_{\max}}{\dot{U}_{\max}}$ is the delay in the relay switching, $U_{\max}$ is the limiting voltage of the Zener diodes D_1 and D_2, and $\dot{U}_{\max}$ is the maximum rate of change of the operational amplifier output voltage.

It follows from prior analysis (see Chapter 4) that if a time delay is present in the linear part of the system, the LPRS will always have a point of intersection with the real axis, and therefore finite-frequency oscillations (chattering) will exist. It also results in the deterioration of the closed-loop performance.

Considering that the equivalent gain k_n is obtained as a finite value, we write the transfer function of the closed-loop differentiator dynamics:

$$W(s) = \frac{k_n e^{-\tau s}/(Ts+1)}{1 + k_n W_l(s)} = \frac{k_n e^{-\tau s}(Ts+1)}{s(Ts+1)^2/K + k_n e^{-\tau s}} \cdot \frac{s}{K}. \qquad (8.5)$$

In formula (8.5), the first multiplier represents the effect of the parasitic dynamics due to the limited bandwidth of the operational amplifiers, and the second multiplier represents the ideal differentiation. The smaller T and τ are, the smaller is the effect of parasitic dynamics, and the higher the frequency of chattering and the higher the value of the equivalent gain. Ideally, if we set $T \to 0, \tau \to 0$, then the differentiator transfer function $W(s) \to \frac{s}{K}$, which is the transfer function of the ideal differentiator. However, one can see that this is impossible due to the limited bandwidth of the operational amplifiers.

It is also worth noting that formula (8.5) relates the average value of the relay output with the input. To obtain the average value of this signal, an additional low-pass filtering has to be applied to the differentiator output. This also contributes to the differentiation error. The issue of low-pass filtering of the discontinuous signal is addressed in the following section devoted to the design of a temperature sensor compensator.

Above, we considered the analog element implementation of the SM differentiator. If the differentiator is realized by means of a digital controller, we can consider only one type of parasitic dynamics, which in this case arises in the form of the processing delay. The transfer function of the linear part in this case is

$$W_l(s) = \frac{e^{-\tau s}}{s}. \tag{8.6}$$

The delay value is not exactly equal to the execution period (cycle) of the controller; the relationship between those two values is more complex. This relationship is addressed in detail in the following chapter devoted to the SM observer. What is important in the current analysis is that in parasitic dynamics exist, and they are manifested as an equivalent time delay. One can see that formula (8.6) can be obtained from (8.4) by setting $T = 0, K = 1$. Therefore, let us obtain the transfer function of the SM differentiator from formula (8.5) assuming that $T = 0, K = 1$:

$$W(s) = \frac{k_n e^{-\tau s}}{s + k_n e^{-\tau s}} \cdot s. \tag{8.7}$$

In formula (8.7), the first multiplier represents the effect of parasitic dynamics due to the processing delay, and the second multiplier represents the ideal differentiation.

The following example illustrates analysis of the performance of the analogue SM differentiator.

Example 8.1. Let the differentiator be implemented as an analog circuit (Fig. 8.1) with the following parameters: $R_1 = R_2 = R_3 = R_4 = R_5 = 100$ kOhm, $C_1 = 10^3$ nF, $f_{\max} = 1$ MHz, $U_{\max} = 10$ V, $\dot{U}_{\max} = 6 \cdot 10^7$ V/s. We calculate the parameters of the transfer function (8.4): $K = 10\,\text{s}^{-1}$, $T = 1.59 \cdot 10^{-7}$ s, $\tau = 1.67 \cdot 10^{-7}$ s. We compute the LPRS for the transfer function (8.4) using formula (2.33). The LPRS is presented in Fig. 8.2.

The frequency of the periodic oscillation in the differentiator loop is the point of intersection of the LPRS and the real axis: $\Omega = 3.33 \cdot 10^6\,\text{s}^{-1}$ ($\sim$530 kHz), and the value of the real part of the LPRS at this point is $\text{Re}J(\Omega) = -1.98 \cdot 10^{-6}$, which provides the value of the equivalent gain $k_n = 2.52 \cdot 10^5$. We compute the frequency response using formula (8.5) and plot the Bode diagram (Fig. 8.3). One can see from the Bode plot that the SM differentiator provides accurate differentiation at lower frequencies of the input signal (subject to subsequent low-pass filtering to get rid of the oscillatory component). Namely, up to frequencies of about 10^6 s^{-1} ($\sim$159 kHz) the accuracy is very high. At higher frequencies of the input signals the accuracy is lower, and the frequency of the input signal cannot exceed the frequency of the self-excited oscillation in the observer loop Ω (in fact, the input frequency should be at least a few times lower than Ω). This example demonstrates the analysis of the performance of the SM differentiator and highlights certain limitations of possible applications of SM differentiators.

8.3 Temperature sensor dynamics compensation via SM application 157

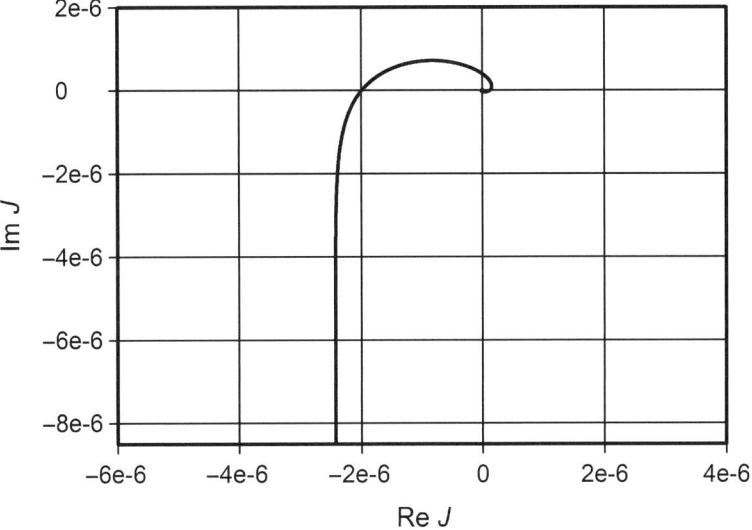

Fig. 8.2. The LPRS of analog differentiator

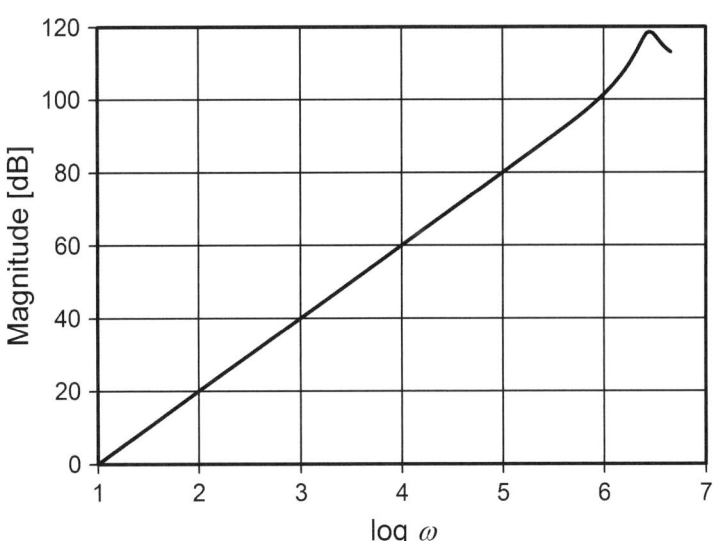

Fig. 8.3. Bode plot of analog differentiator

8.3 Temperature sensor dynamics compensation via SM application

Precise temperature measurements in the combustion chambers of modern utility boilers, incinerators, and Claus furnaces in various transient processes an important technical problem. Accurate temperature measurement are allows one to sense relatively fast fluctuations of this important internal process parameter and implement more sophisticated control schemes to achieve better control. The same also applies to the measurement of water and steam temperature in utility boilers. However, unless pyrometer measurements are used (which is an expensive option), the dynamics of the sensor due to the heat transfer in the thermowell and sensor material does not allow one to read the process temperature instantaneously. The lag may be significant, preventing the device from sensing fast temperature changes and fluctuations in a transient mode. The combined thermowell-sensor time constant may range from a few seconds to a few minutes.

A possible means of compensating for the sensor lag effect is the application of the transfer function "inversion" principle based on SM. This application is a special type of the SM observer, which allows one to obtain a variable that cannot be measured directly, via matching the measured variables, subject to the availability of the sensor dynamic model. This compensation can be used within the electronics of temperature transmitters. The theory of SM observers that can be utilized for this purpose is presented in [38, 97]. However, this theory does not provide any non-ideal quantitative characteristics of a SM observer in a transient mode, for example under harmonic excitation. It assumes that once the SM is established, an ideal observation occurs.

With the LPRS method, which considers parasitic dynamics and their effects, including the "chattering" phenomenon and non-ideal response to external inputs and disturbances, one is able to do a quantitative assessment of compensator quality in various transient modes.

Let the dynamics of the temperature sensor be described by the following transfer function [35] (this is an approximate model, and for that reason the device being considered would provide only partial compensation),

$$\frac{T_s(s)}{T_c(s)} = W_s(s) = \frac{1}{(T_1 s + 1)(T_2 s + 1)} \tag{8.8}$$

where T_1 is the time constant due to the thermowell heat transfer, T_2 is the sensor time constant, $T_s(s)$ is the temperature signal provided by the sensor (in the Laplace domain), and $T_c(s)$ is the true combustion temperature (in the Laplace domain).

Let us design a SM compensator using the same idea as that of SM observer design [97]. The SM observer includes a model, which has the same dynamics as the dynamics of the thermowell-sensor. This model has a discontinuous control input that enforces the output of the model to match the measured temperature. This is achieved via discontinuous SM control, which acts to eliminate the mismatch between the measured temperature provided by the sensor and the output of the model

$$\ddot{\hat{T}}_s = \frac{1}{T_1 T_2} \left[u - (T_1 + T_2)\dot{\hat{T}}_s - \hat{T}_s \right] \tag{8.9}$$

8.3 Temperature sensor dynamics compensation via SM application

$$u = \begin{cases} +c \text{ if } \sigma \geq 0 \\ -c \text{ if } \sigma < 0, \end{cases} \tag{8.10}$$

$$\sigma = T_s - \hat{T}_s \tag{8.11}$$

where $\hat{T}_s$ is the output of the compensator's internal model (which is supposed to match the sensor reading T_s), u is the discontinuous control applied to the compensator internal model, c is the amplitude of this control, and σ is the error signal for the SM compensator (the mismatch between the sensor reading and the compensator internal model output).

The mode that occurs in the SM compensator can be characterized as an asymptotic second-order SM [1, 40], since the relative degree of the plant given by equation (8.9) is *two*. The variable that represents an estimate of the true combustion temperature is the equivalent control $u_{eq}(t)$, which is the averaged discontinuous control $u(t)$ subject to the existence of ideal SM.

With the compensator model given by equations (8.9)–(8.11), the mode that occurs in the compensator closed loop is an ideal SM revealed as infinite frequency oscillations. Obviously, such a mode cannot exist in a real system. Due to the inevitable existence of some parasitic dynamics not accounted for in the compensator model, the real mode has high but finite frequency oscillations. This can be analyzed with the use of the LPRS method if the model of these parasitic dynamics is available. A relatively precise model of parasitic dynamics can be derived from the characteristics of the components of the SM compensator (limited bandwidth of the operational amplifiers, delays and hysteresis in the switching elements, etc.). For the purpose of the current analysis which is mostly illustrative, assume that the parasitic dynamics of the SM compensator is manifested as the hysteresis of the relay function given by formula (8.10). In this case the control function can be rewritten as:

$$u = \begin{cases} +c \text{ if } \sigma = f_0 - y \geq b \text{ or } \sigma > -b, u(t-0) = c \\ -c \text{ if } \sigma = f_0 - y \leq -b \text{ or } \sigma < b, u(t-0) = -c, \end{cases} \tag{8.12}$$

where b is the hysteresis value (half of the total hysteresis value) of the relay, and $u(t-0)$ is the control value at the time immediately preceding the current time.

In the system (8.9), (8.11), (8.12), high-frequency self-excited oscillations occur, which we refer to as a real SM. As a result of this, the averaged control is now slightly different from the equivalent control. The averaged control that is used as an assessment of the combustion temperature can be obtained by low-pass filtering of the control $u(t)$. For that purpose, a low-pass filter (i.e., second-order filter) must be included in the compensator model as follows:

$$\hat{T}_c(s) = \frac{1}{T_3^2 s^2 + T_4 s + 1} u(s). \tag{8.13}$$

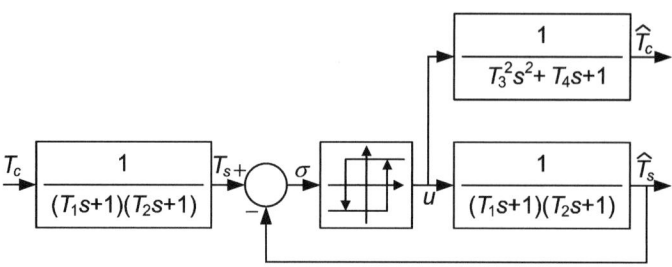

Fig. 8.4. Sensor dynamics SM compensator

The resulting dynamical model of the senor and compensator can now be represented as a block diagram (Fig. 8.4). It is worth noting here that the SM compensator is essentially a relay servo system. Therefore, all applicable methods of analysis that are used for relay systems can also be applied here.

8.4 Analysis of the sliding mode compensator

Let us carry out two types of analysis of compensator dynamics: under the ideal SM assumption and under the non-ideal SM hypothesis. At first, we assume that the control is given by (8.10). Then ideal SM occurs [1]. The averaged control u_0 in this case is equal to the equivalent control u_{eq}. The latter can be obtained by replacing the original relay nonlinearity (8.10) with infinite gain (see Chapter 4 for details):

$$u_0 = K\sigma_0, \quad K \to \infty. \tag{8.14}$$

The subscript "0" in (8.14) refers to the averaged values. Therefore, for the averaged motion, (8.10) and (8.11) can be rewritten as follows,

$$u_0 = K\left(T_s - \hat{T}_{s0}\right), \quad K \to \infty, \tag{8.15}$$

where $\hat{T}_{s0}$ is the value $\hat{T}_s$ averaged over the period of chattering. The equations of averaged motions can be derived from (8.15) as follows. Because u_0 is always finite, the only way for (8.15) to hold is for the variables T_s and $\hat{T}_{s0}$ to be equal. This in turn leads to the following equality:

$$\frac{u_{eq}(s)}{\hat{T}_{s0}(s)} = \frac{u_0(s)}{\hat{T}_{s0}(s)} = (T_1 s + 1)(T_2 s + 1). \tag{8.16}$$

Therefore, $u_{eq} = T_c$. This conclusion is in agreement with the theory of SM observation. However, this represents an ideal situation which cannot exist in real applications.

Let us carry out a similar analysis, but consider the hysteretic character of the relay nonlinearity. At first, the LPRS of the linear part of the system must be computed through formula (2.19), where $W_l(s) = W_s(s) = \frac{1}{(T_1 s+1)(T_2 s+1)}$.

8.4 Analysis of the sliding mode compensator

The series (2.19) converges very quickly and can be used for calculations. Alternatively, for the $W_l(s)$ given by the second-order transfer function, the LPRS has an analytical formula

$$J(\omega) = 0.5[1 - \frac{T_1}{T_1-T_2}\alpha_1 \operatorname{csch} \alpha_1 - \frac{T_2}{T_2-T_1}\alpha_2 \operatorname{csch} \alpha_2]$$
$$-j0.25\frac{\pi}{T_1-T_2}[T_1 \tanh \frac{\alpha_1}{2} - T_2 \tanh \frac{\alpha_2}{2}], \quad (8.17)$$

where $\alpha_1 = \frac{\pi}{T_1 \omega}$, $\alpha_2 = \frac{\pi}{T_2 \omega}$.

Once the LPRS is computed, equation (2.3) must be solved, from which the frequency of periodic motions in the SM compensator can be found. After that, the equivalent gain of the relay, which relates the averaged values of the input to and output of the relay, must be computed as per (2.4). With the equivalent gain value available, one can write the equations of the averaged motions in the compensator:

$$\ddot{\hat{T}}_{s0} = \frac{1}{T_1 T_2}\left[u_0 - (T_1+T_2)\dot{\hat{T}}_{s0} - \hat{T}_{s0}\right] \quad (8.18)$$

$$u_0 = k_n \sigma_0 \quad (8.19)$$

$$\sigma_0 = T_s - \hat{T}_{s0} \quad (8.20)$$

One can see that the equations of the averaged motions are linear, which occurs due to the chatter smoothing phenomenon. Now by closing the open-loop equations, we find the transfer function of the compensator:

$$W_c(s) = \frac{u_0(s)}{\hat{T}_{s0}(s)} = (T_1 s + 1)(T_2 s + 1)\frac{1}{1 + \frac{1}{k_n}(T_1 s + 1)(T_2 s + 1)} \quad (8.21)$$

Comparing (8.21) and (8.16), one can see that the real transfer function (8.21) of the compensator has an additional factor, given by the fraction in (8.21), in comparison to the transfer function of the ideal compensator (8.16). If $k_n \to \infty$, this factor tends to unity and the real transfer function becomes equal to the ideal one. If the equivalent gain is a high but finite value, the factor in (8.21) represented by the fraction is a low-pass filter, which reduces the quality of compensation at high frequencies. Therefore, the bandwidth of the compensator depends on the value of the hysteresis. The value of the hysteresis b at given T_1 and T_2 determines both the frequency of periodic motion and the value of the equivalent gain of the relay. This is illustrated below in the example of compensator design. The low-pass filter, which serves the purpose of suppressing the periodic component in signal $u(t)$ in order to obtain $\hat{T}_{s0}$, must be designed using conventional frequency-domain techniques and techniques for separating two signals of frequencies that differ by a factor of more than 100.

8.5 An example of compensator design

Let us design the SM compensator for the sensor, the dynamics of which are presented by the following two time constants: $T_1 = 5$ s, $T_2 = 2$ s. Let the frequency range of possible temperature variations be $\omega \in [0; 0.2\,\text{Hz}]$. Therefore, the required bandwidth of the designed compensator must be at least the same. Denote $W_s(s) = \frac{1}{(5s+1)(2s+1)}$. Then the LPRS $J(\omega)$ corresponding to this transfer function can be computed as in (8.17). It is presented in Fig. 8.5.

If the hysteresis $b = 0$, self-excited oscillations of infinite frequency would occur in the SM compensator loop of the system in Fig. 8.4. Assuming $c = 10$ and a small value of hysteresis $b = 5 \cdot 10^{-10}$ (which represents parasitic dynamics and is absolutely necessary in the model to obtain a finite-frequency solution), we calculate $-\frac{\pi b}{4c} = -3.93 \cdot 10^{-11}$. According to (2.3), we compute the frequency of self-excited oscillations in the SM observer loop: $\Omega = 1218$ rad/s. The real part of the LPRS at this frequency is $\text{Re}J(\Omega) = -5.54 \cdot 10^{-8}$. Now we calculate the equivalent gain of the relay as per (2.4): $k_n = 9.02 \times 10^6$. The dynamical model of the averaged motions in the sensor-compensator system is given in Fig. 8.6. We design the low-pass filter to filter out the frequency of self-excited oscillations.

Consider two frequencies: the upper frequency of the required bandwidth $2\pi \cdot 0.2$ Hz$= 1.256$ rad/s and the frequency of self-excited oscillations $\Omega = 1218$ rad/s. Because the two frequencies are far from each other, we select a two-pole Butterworth filter as the low-pass filter. We select the natural frequency of the filter to be eight times higher than the upper frequency of the bandwidth $\omega_n =$

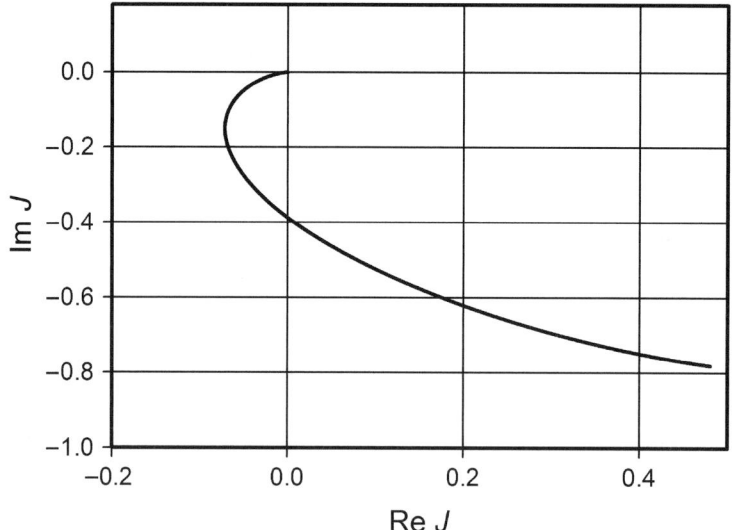

Fig. 8.5. The LPRS $J(\omega)$ for transfer function $W(s)$

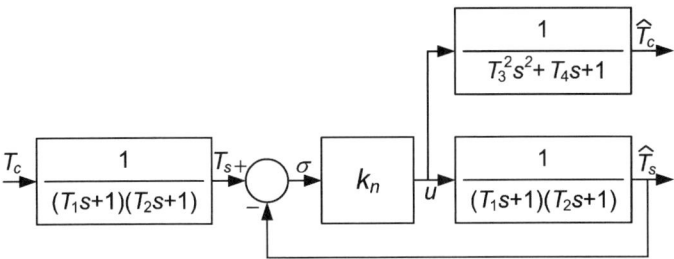

Fig. 8.6. Dynamics of averaged motions in SM compensator

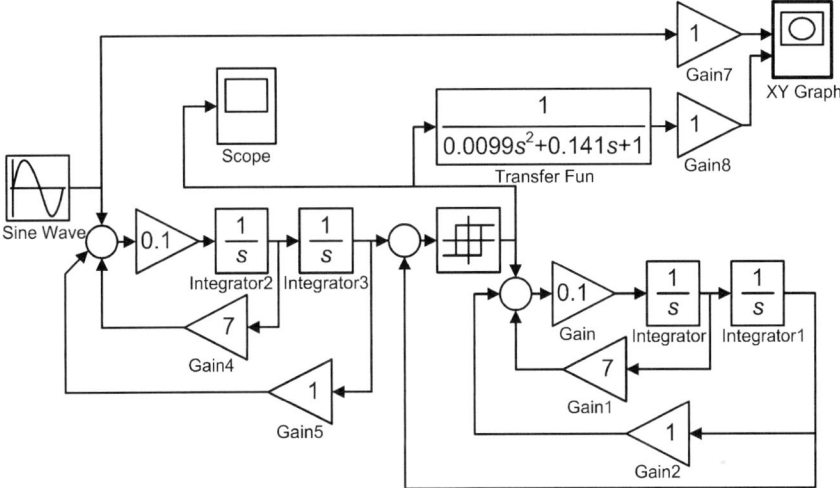

Fig. 8.7. Simulink model of sensor-compensator dynamics

10.05 rad/s and calculate T_3 and T_4 as follows: $T_3 = 1/\omega_n = 0.0995\,\text{s}$, $T_4 = \sqrt{2}/\omega_n = 0.141\,\text{s}$. Now let us run simulations of the designed SM compensator. The Simulink model of the compensator is presented in Fig. 8.7.

We run a few different simulations using this model. The output of the sliding mode controller (relay) is presented in Fig. 8.8, which shows that self-excited oscillations of the predicted frequency indeed exist in the SM compensator loop. The response to a harmonic input $T_c(t) = \sin 0.5t$ is presented in Fig. 8.9. In this figure, the horizontal axis represents the actual temperature $T_c(t)$, and the vertical axis is the estimated temperature $\hat{T}_c$. There is a small phase lag between the two signals, which is mainly due to the phase lag introduced by the low-pass filter. The use of a higher-order filter would improve

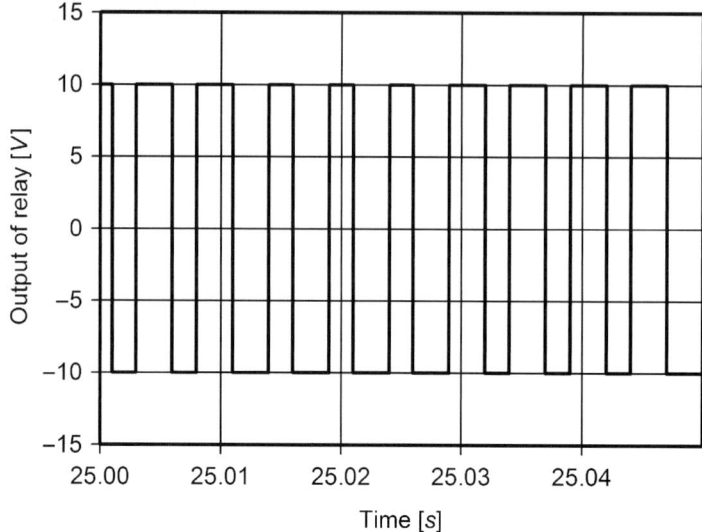

Fig. 8.8. Output of the relay (scope of the Simulink diagram)

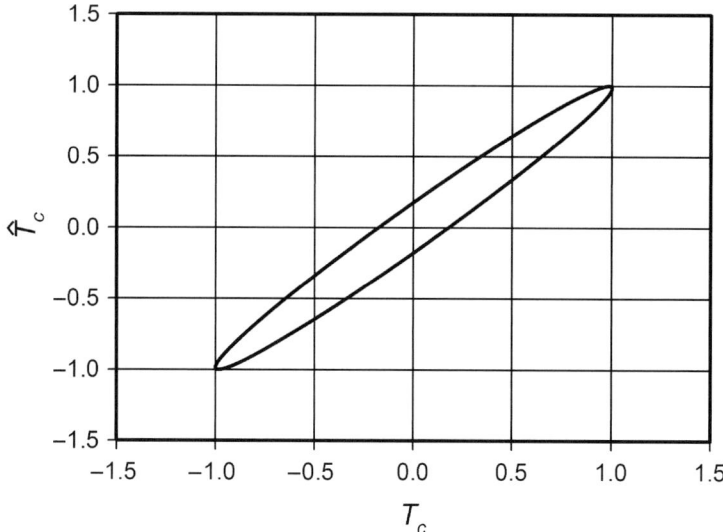

Fig. 8.9. True temperature–observed temperature relationship

the quality of the compensator. The response of the compensator to the combination of a ramp input and a harmonic input is presented in Fig. 8.10. One can see from this figure that the output of the compensator $\hat{T}_c$ tracks the true temperature $T_c(t)$ very closely.

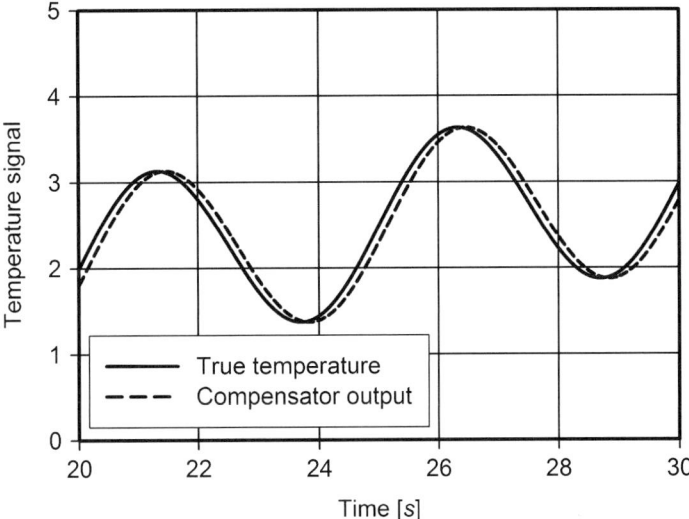

Fig. 8.10. True temperature and observed temperature: harmonic and ramp change

8.6 Conclusions

We consider a sliding mode differentiator and a SM compensator for temperature sensor dynamics. The differentiator can be used for taking the derivative of various signals. The compensator is capable of restoring the original temperature variations, which otherwise are lagged. It can be used as a part of a temperature transmitter improving its dynamic response.

The compensation is based on the sliding mode observation principle. The sliding mode is generated in the compensator loop, which includes the model of either the integrator for the former or the sensor dynamics for the latter and the SM controller. We show that if ideal sliding mode occurs, then the equivalent control is equal to the observed variable. However, due to the inevitable presence of parasitic dynamics, the ideal SM cannot occur, and a real SM occurs instead. We show that in this case, the true derivative or the true temperature are approximately equal to the averaged control signal. Those conclusions are illustrated with examples of design and simulations.

9
Analysis of sliding mode observers

9.1 The SM observer as a relay servo system

The idea of using a dynamical system to obtain estimates of the system states from measurable system variables was first proposed by Luenberger [69]. This dynamical system is known as the *observer*. In this approach, the observer dynamical system is driven by the control and by the difference between the output of the observer and the output of the plant. Ideally, this difference is zero, which indicates that the state estimates generated by the observer are equal to the states of the plant.

SM can also be used for the purpose of observation if generated in the observer dynamical system, and the system is designed in such a way that the difference between the output of the observer and the output of the plant becomes a *sliding variable*. The control should be designed to provide the existence of SM in the observer dynamical system. This is the main idea of the SM observer.

SM observers are analyzed in a number of publications (see, for example, respective chapters of texbooks [38, 97], papers [82, 101], and recent tutorials [10] and [39]). However, only the ideal SM in the observer dynamical system is analyzed in those works (as well as in the publications referenced above). For that reason, in a steady mode, the observation is done with zero error. The nonzero observation error only appears in the transients while the process is converging. After the process has converged (this time may be infinite, though, and for that reason some observers are called *asymptotic observers*), the observation error becomes zero and ideal observation occurs. Another feature of the traditional analysis is that SM observers are always analyzed as stabilization systems but not as servo systems.

However, no ideal SM exists in any real application. Parasitic dynamics always exist along with the principal dynamics. Even if we consider an observer in which the control is realized by means of a digital processor, the processor itself introduces a delay into the observer loop. In addition to that, the sensors that measure the variables of the plant also contribute to the error of observation, despite that fact that they are outside the observer loop, due to a mismatch between the observer model, which does not include the parasitic dynamics, and the plant, which does include these dynamics. It was shown in Chapter 4 that due to the presence of parasitic dynamics, not only does chattering occur but also the input-output properties of the SM system differ from those of the reduced-order system model. Naturally, we can expect the same effect in the SM observers, too.

Consider an n-dimensional version of the observer proposed by Utkin [97]. Let the linear plant, the states of which are supposed to be observed, be the n-th order dynamical system

$$\dot{\mathbf{x}} = \mathbf{A}\mathbf{x} + \mathbf{B}u \tag{9.1}$$

$$y = \mathbf{C}\mathbf{x}, \tag{9.2}$$

where $\mathbf{x} \in R^n$ is a state vector, $y \in R^1$ is the measurable system output, and $\mathbf{A} \in R^{n \times n}, \mathbf{B} \in R^{n \times 1}$, and $\mathbf{C} \in R^{1 \times n}$ are matrices. The pair $(\mathbf{C},\mathbf{A})$ is assumed to be observable.

The SM observer can be designed in the same form as the original system (9.1), (9.2) with the addition of an *output injection* being, in fact, an observer correction input which depends on the error between the output of the observer and the output of the plant (the system to be observed):

$$\dot{\hat{\mathbf{x}}} = \mathbf{A}\hat{\mathbf{x}} + \mathbf{B}u + \mathbf{L}\,\text{sign}(y - \hat{y}) \tag{9.3}$$

$$\hat{y} = \mathbf{C}\hat{\mathbf{x}} \tag{9.4}$$

where $\hat{\mathbf{x}}$ is an estimate of the system state vector, $\hat{y}$ is an estimate of the system output, and $\mathbf{L} \in R^{n \times 1}$ is a gain matrix.

We denote the sliding variable as follows:

$$\sigma = y - \hat{y}. \tag{9.5}$$

The elements of $\mathbf{L}$ must ensure the reachability condition of the SM and stability of the reduced SM dynamics. It is shown in [38, 97] that the matrix $\mathbf{L}$ can be selected to guarantee convergence of the sliding variable σ to zero in finite time and asymptotic convergence of the estimation error for the system variables. We assume that the conditions for the existence of SM in the observer dynamics are satisfied. The subject of analysis of the current chapter is the effect of the inevitably present parasitic dynamics in any practical realization of an observer.

Let us consider the observer to be a dynamical system that has two inputs and one output. One of those two inputs $y(t)$ must be followed (tracked) by the observer output as precisely as possible. The other input $u(t)$ can be treated as a feedforward. Therefore, it makes sense to consider the observer as a relay servo system (Fig. 9.1).

With this representation, we are now able to apply the LPRS approach presented earlier to analysis of the SM observer performance. Yet, it is worth

9.1 The SM observer as a relay servo system 169

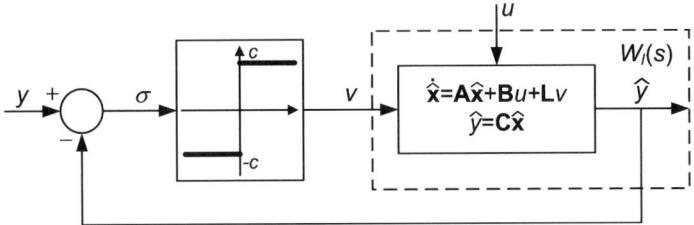

Fig. 9.1. Relay servo system representation of the SM observer

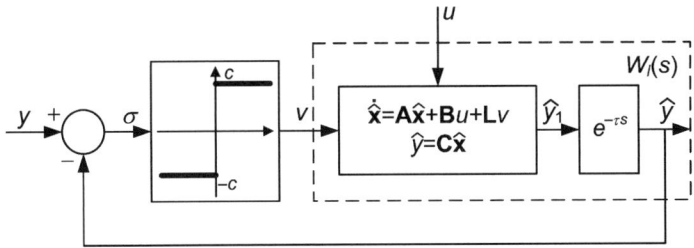

Fig. 9.2. Relay servo system representation of the SM observer with parasitic dynamics

mentioning that a SM observer is designed so that, theoretically, in the system in Fig. 9.1, an ideal SM occurs. As a result, in a steady state, the variables $\hat{y}(t)$ and $y(t)$ are always equal, and the state estimates $\hat{\mathbf{x}}(t)$ are equal to the plant states $\mathbf{x}(t)$. To be able to analyze the dynamic performance of the SM observers, we need to determine what kind of parasitic dynamics are present in the system. Accounting for parasitic dynamics results in a system model in which the ideal sliding mode no longer exists, and observation accuracy is not ideal.

Considering a digital realization of the SM observer, we note that the parasitic dynamics of the SM observer come from the delay in the output calculation [6]. We show now that in a relay feedback system, the digital realization of the observer equations is manifested as a delay. Consider only the calculations that are done in the observer loop. Suppose that the computing of the observer output begins from the calculation of the sliding variable $\sigma(t)$, followed by the calculation of the discontinuous control $v(t)$, and so on up to the observer output $\hat{y}(t)$. All these calculations are done over one execution period of the controller. Therefore, for the calculation of the sliding variable, the value of the observer output calculated on the previous execution period is used. This is equivalent to introducing the time delay in the observer loop (Fig. 9.2). We note that the execution period for computing $\hat{y}(t)$ and the sampling rate for measuring $y(t)$ are two different notions. However, for the purpose of our analysis, we can assume that those two values are equal. This is a simplified approach. In reality, there may be different combinations of sampling rate and execution period. The results also depend on the algorithm of numeric integration of the differential equations.

Let us proceed from the following assumption. Because in the absence of parasitic dynamics the SM is realized ideally, the digital implementation exhibits chattering with the period equal to two execution periods of the algorithm (see [77], for example). On the other hand, the system with a time delay (Fig. 9.2) exhibits the same kind of motion. Obviously, by varying the time delay, the frequency of chattering in the second case can be tuned to be equal to the frequency of chattering in the first case. If we assume that the solution of system (9.3), (9.4) is obtained exactly at the digital realization of the SM observer, then the two SM observer dynamics, namely, the digital realization and the time delay realization, are equivalent. In other words, if we know the execution period, we can find a certain equivalent time delay such that the dynamics of these two representations of the SM observer are equivalent. A similar approach to analysis of discrete-time SM systems is proposed in [44]. Therefore, the use of time delay in the model represents a certain generic approach to the evaluation of SM observer performance.

With this representation, we can analyze observer performance in terms of the response of the relay servo system (Fig. 9.2) to two inputs: u and y. This is a complex task, however, which can be fulfilled by application of the LPRS method, which is well suited for input-output analysis of relay systems.

9.2 SM observer performance analysis and characteristics

With the representation of the SM observer as a relay servo system, we can formulate certain performance measures of the observer. In our analysis, we have to consider certain properties of this servo system typical of the observers only. The main feature of the observer input-output dynamics is that there are two different inputs to the system that are not independent (Fig. 9.3).

From the function of the observer it follows that, apart from the initial transient time, the values of $y(t)$ and of $\hat{y}(t)$ are very close. The variable $y(t)$ is a result of the propagation of the input $u(t)$ through the dynamics of the plant. On the other hand, if we assume that the model of the observer ideally matches the model of the plant, then by setting $v(t) \equiv 0$, we can conclude that $\hat{y}(t) = y(t)$, which essentially means that the discontinuous control $v(t)$ is necessary to compensate for the mismatch between the dynamics of the plant and the dynamics of the observer model.

Because the values of $y(t)$ and of $\hat{y}(t)$ are close, with the value of $\hat{y}(t)$ alternately slightly above or slightly below $y(t)$, the control $v(t)$ is almost equally spaced, and the motion of the observer state variables is driven mainly

9.2 SM observer performance analysis and characteristics

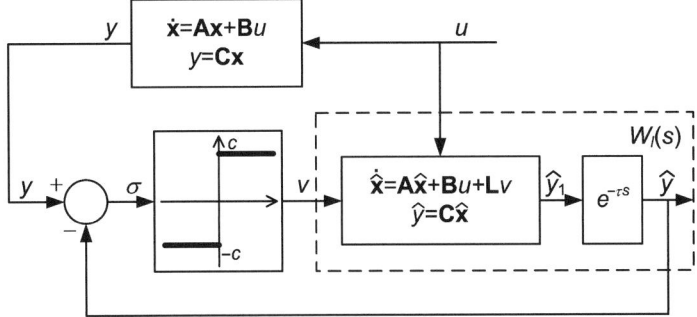

Fig. 9.3. Plant and observer model

by the control $u(t)$ ($v(t)$ applies a corrective action to match the plant and observer outputs).

At first, consider a possible methodology of analysis of the observer dynamical system given in Fig. 9.2. From the LPRS theory, we know that the averaged forced motions in the system in Fig. 9.2 can be analyzed using the concepts of the equivalent gain of the relay and the linearized model, which can be obtained from the original model by replacing the relay function with the equivalent gain. This is the general approach. The linear part of the system for the LPRS analysis is the dynamics of the observer model and the parasitic dynamics (time delay) marked in the diagram in Fig. 9.2 with the dashed line. This model can also be represented by the transfer function $W_l(s)$ from the control $v(t)$ to the output $\hat{y}(t)$.

The methodology of input-output analysis of the dynamics given in Fig. 9.3 does not differ much from other applications of the LPRS method. However, the observer analysis has its specifics, due to unknown value of the equivalent time delay. As a result, analysis is done as before through the following steps: (a) identification of the equivalent time delay of the continuous-time model of the observer by matching the frequencies of chattering, where the frequency of chattering of the continuous-time system is computed from (2.3), where $b = 0$, with the LPRS computed using (2.12), (2.19), or (2.32); (b) computation of the LPRS for the linear part of the relay servo system (corresponding to $W_l(s)$); (c) computation of the equivalent gain value using formula (2.4); and (d) replacement of the relay with the equivalent gain and analysis of the linearized observer dynamics. The linearized plant-observer dynamics can be represented as the diagram Fig. 9.4.

In Fig. 9.4, subscript "0" is used to indicate the average over the period of chattering variables.

We have thus obtained a linear model of the plant and the observer, from which various characteristics of the observer accuracy can be derived. Those characteristics are different from the frequency and the amplitude of chattering that can be obtained through the relay feedback representation of the observer (Fig. 9.2). If, for example, We follow the conventional approach to

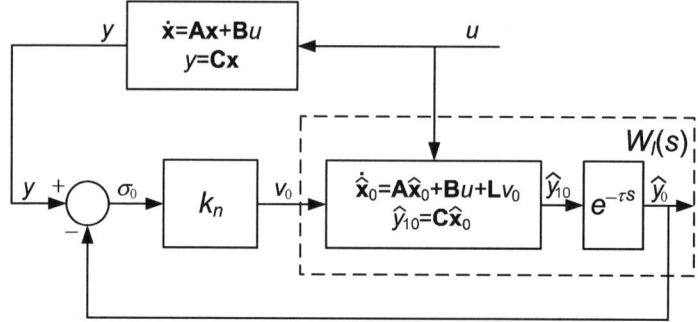

Fig. 9.4. Linearized model of plant and observer

servo systems analysis, we can formulate a dynamical accuracy criterion as a frequency response of the error signal $\sigma(t)$ to the harmonic excitation $u(t)$ of variable frequency. This characteristic can be presented as magnitude and phase responses

$$M = 20 \lg |W_{u-\sigma}(j\omega)| \tag{9.6}$$

$$\varphi = \arg W_{u-\sigma}(j\omega) \tag{9.7}$$

where M is the magnitude response, φ is the phase response, and $W_{u-\sigma}(s)$ is the transfer function from $u(t)$ to $\sigma(t)$ given by

$$W_{u-\sigma}(s) = \mathbf{C}(s\mathbf{I} - \mathbf{A})^{-1}\mathbf{B} \frac{1 - e^{-\tau s}}{1 + k_n \mathbf{C}(s\mathbf{I} - \mathbf{A})^{-1}\mathbf{L}e^{-\tau s}}. \tag{9.8}$$

Such characteristics as the bandwidth, resonant frequencies, and others can be easily obtained from the frequency response of the linearized plant-observer model, too.

9.3 Example of SM observer performance analysis

Let the plant be the second-order system

$$\dot{\mathbf{x}} = \mathbf{A}\mathbf{x} + \mathbf{B}u \tag{9.9}$$

$$y = \mathbf{C}\mathbf{x}, \tag{9.10}$$

where $\mathbf{A} = \begin{bmatrix} 0 & 1 \\ -1 & -3 \end{bmatrix}$, $\mathbf{B} = \begin{bmatrix} 0 \\ 1 \end{bmatrix}$, $\mathbf{C} = \begin{bmatrix} 1 & 1 \end{bmatrix}$, and the observer dynamics are as follows,

$$\dot{\hat{\mathbf{x}}} = \mathbf{A}\hat{\mathbf{x}} + \mathbf{B}u + \mathbf{L}\,\mathrm{sign}(y - \hat{y}) \tag{9.11}$$

$$\hat{y} = \mathbf{C}\hat{\mathbf{x}}, \tag{9.12}$$

where $\mathbf{L} = \begin{bmatrix} 1 \\ 4 \end{bmatrix}$.

9.3 Example of SM observer performance analysis

Let the equivalent time delay be $\tau = 0.01\text{s}$. The corresponding execution period that causes such delay is determined below. Assume that the input to the system is a harmonic oscillation of frequency ω that can be varied: $u(t) = \sin(\omega t)$.

We write an expression for the transfer function of the linear part in the relay servo dynamics of the observer:

$$W_l(s) = \mathbf{C}(s\mathbf{I} - \mathbf{A})^{-1}\mathbf{L}e^{-\tau s} = \frac{5s+6}{s^2+3s+1}e^{-0.01s}$$

$$= \left[\frac{\frac{25+3\sqrt{5}}{10}}{s+\frac{3+\sqrt{5}}{2}} + \frac{\frac{25-3\sqrt{5}}{10}}{s+\frac{3-\sqrt{5}}{2}}\right]e^{-0.01s}$$

$$= \frac{1.2112}{0.3819s+1}e^{-0.01s} + \frac{4.7888}{2.6180s+1}e^{-0.01s}. \tag{9.13}$$

The transfer function in (9.13) is represented in terms of partial fractions, which allows for the use of the analytical expression for the LPRS. Now we compute the LPRS corresponding to the transfer function (9.13) using the formula for the first-order plus dead-time dynamics (2.64) and Theorem 2.1:

$$J(\omega) = \frac{K}{2}(1 - \alpha e^\gamma \operatorname{csch}\alpha) + j\frac{\pi}{4}K\left(\frac{2e^{-\alpha}e^\gamma}{1+e^{-\alpha}} - 1\right)$$

where $\alpha = \frac{\pi}{T\omega}$ and $\gamma = \frac{\tau}{T}$, and τ is the time delay and T is the time constant.

The LPRS of the observer dynamics is presented in Fig. 9.5.

We calculate the frequency of chattering using formula (2.3) assuming zero hysteresis ($b = 0$):

$$\Omega = 158.48 s^{-1}.$$

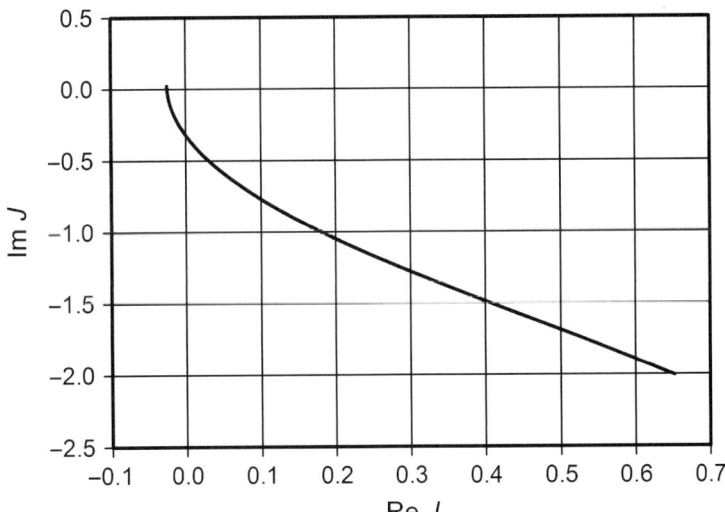

Fig. 9.5. The LPRS of the SM observer linear part

174   9 Analysis of sliding mode observers

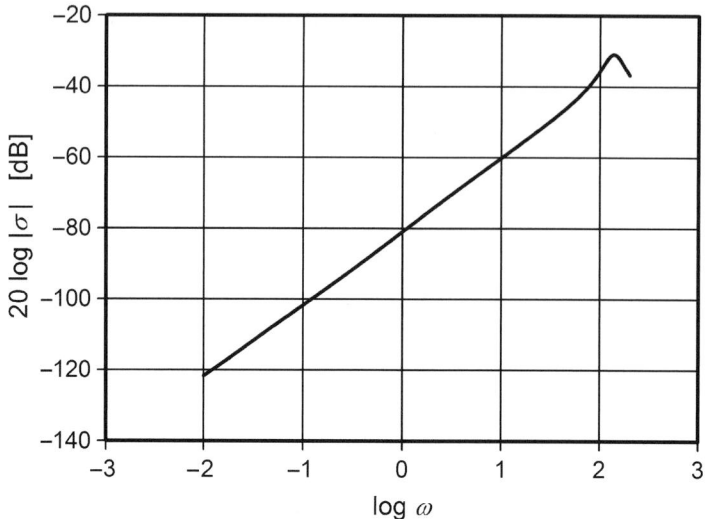

Fig. 9.6. Output error magnitude versus frequency

This corresponds to an execution period equal to $\pi/\Omega = 0.0198$ s, which illustrates the correspondence between the execution period and the equivalent delay. Using formula (2.4), we calculate the equivalent gain of the relay:

$$k_n = 20.08.$$

Now we can compute the observation error for various frequencies of the input signal as the magnitude of the transfer function (9.8). The plot providing the output error versus the frequency of the input signal is given in Fig. 9.6. From this plot, we can also derive some other characteristics of the analyzed SM observer. For example, the resonant frequency of the observer characteristic is about 138 s^{-1}, the observer bandwidth measured at the level of 1% observation error is about 76 s^{-1}, and maximum observation error is about 2.85%. It should be noted that the frequency-domain input-output analysis in a SM system makes sense only for frequencies below the frequency of chattering. For that reason, the observation error is computed only in this frequency range.

The observation error can also be computed for x_1 and x_2 using the same methodology

$$W_{u-x1}(s) = \mathbf{C}(s\mathbf{I} - \mathbf{A})^{-1}\mathbf{B}\frac{1 - e^{-\tau s}}{1 + k_n \mathbf{C}(s\mathbf{I} - \mathbf{A})^{-1}\mathbf{L}e^{-\tau s}}\mathbf{C}_1(s\mathbf{I} - \mathbf{A})^{-1}\mathbf{L} \quad (9.14)$$

where $\mathbf{C}_1 = \begin{bmatrix} 1 & 0 \end{bmatrix}$;

$$W_{u-x2}(s) = \mathbf{C}(s\mathbf{I} - \mathbf{A})^{-1}\mathbf{B}\frac{1 - e^{-\tau s}}{1 + k_n \mathbf{C}(s\mathbf{I} - \mathbf{A})^{-1}\mathbf{L}e^{-\tau s}}\mathbf{C}_2(s\mathbf{I} - \mathbf{A})^{-1}\mathbf{L} \quad (9.15)$$

where $\mathbf{C}_2 = \begin{bmatrix} 0 & 1 \end{bmatrix}$.

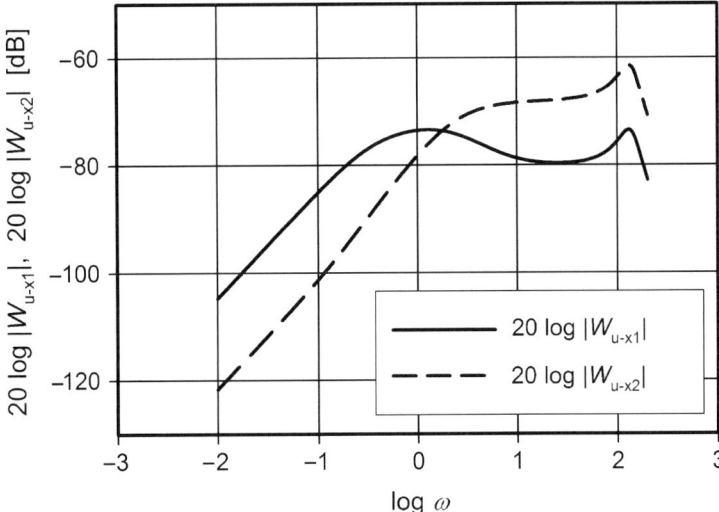

Fig. 9.7. Observation error of x_1 (lower plot) and x_2 (upper plot) versus frequency

The observation error for the variables x_1 and x_2 is plotted against the frequency of the input signal in Fig. 9.7. Computer simulations qualitatively match very well the results obtained analytically. However, the numerical results may differ from the analytical results, depending on the used method of integration, step value, execution order of the model, etc.

9.4 Conclusions

We consider LPRS-based analysis of the SM observer. We show that due to the inevitable presence of parasitic dynamics, an observation error always takes place. We uncover the nature of the parasitic dynamics in the observation algorithm. We show that the parasitic dynamics reveal themselves as a delay in the observer loop. We demonstrate that the delay can be identified by matching the frequency of chattering in the original discrete-time system and in the equivalent continuous-time system having a delay. We provide methodology of subsequent analysis of the observer performance in the frequency domain.

10
Appendix

10.1 The LPRS derivation for a non-integrating linear part

Let the plant be
$$\dot{\mathbf{x}} = \mathbf{A}x + \mathbf{B}u \\ y = \mathbf{C}\mathbf{x} \tag{10.1}$$

and the control
$$u = \begin{cases} +1 \text{ if } \sigma = f_0 - y \geq b & \text{or} \quad \sigma > -b,\ u(t-) = 1 \\ -1 \text{ if } \sigma = f_0 - y \leq -b & \text{or} \quad \sigma < b,\ u(t-) = -1 \end{cases}$$

where $\mathbf{A} \in R^{n \times n}$, $\mathbf{B} \in R^{n \times 1}$, $\mathbf{C} \in R^{1 \times n}$ are matrices, and $\mathbf{A}$ is nonsingular. The solution for the constant control u is
$$\mathbf{x}(t) = e^{\mathbf{A}t}\mathbf{x}(0) + \mathbf{A}^{-1}(e^{\mathbf{A}t} - \mathbf{I})\mathbf{B}u.$$

A fixed point of the Poincaré return map for asymmetric periodic motion with positive pulse length θ_1 and negative pulse length θ_2 and the unity amplitude (the LPRS does not depend on the control amplitude) is given by:

$$\eta_p = e^{\mathbf{A}\theta_1}\rho_p + \mathbf{A}^{-1}(e^{\mathbf{A}\theta_1} - \mathbf{I})\mathbf{B}, \tag{10.2}$$

$$\rho_p = e^{\mathbf{A}\theta_2}\eta_p - \mathbf{A}^{-1}(e^{\mathbf{A}\theta_2} - \mathbf{I})\mathbf{B}, \tag{10.3}$$

where $\rho_p = \mathbf{x}(0) = \mathbf{x}(T)$, $\eta_p = \mathbf{x}(\theta_1)$. Suppose θ_1 and θ_2 are known. Then we can solve (10.2) and (10.3) for ρ_p and η_p. We substitute (10.3) for ρ_p in (10.2),

$$\eta_p = e^{\mathbf{A}\theta_1}\left[e^{\mathbf{A}\theta_2}\eta_p - \mathbf{A}^{-1}(e^{\mathbf{A}\theta_2} - \mathbf{I})\mathbf{B}\right] + \mathbf{A}^{-1}(e^{\mathbf{A}\theta_1} - \mathbf{I})\mathbf{B},$$

which leads to

$$\eta_p = e^{\mathbf{A}(\theta_1+\theta_2)}\eta_p - e^{\mathbf{A}\theta_1}\mathbf{A}^{-1}(e^{\mathbf{A}\theta_2} - \mathbf{I})\mathbf{B} + \mathbf{A}^{-1}(e^{\mathbf{A}\theta_1} - \mathbf{I})\mathbf{B}.$$

We regroup the above equation.

$$(\mathbf{I} - e^{\mathbf{A}(\theta_1+\theta_2)})\eta_p = \left[-e^{\mathbf{A}\theta_1}\mathbf{A}^{-1}(e^{\mathbf{A}\theta_2} - \mathbf{I}) + \mathbf{A}^{-1}(e^{\mathbf{A}\theta_1} - \mathbf{I})\right]\mathbf{B}.$$

We express η_p from the above.

$$\eta_p = (\mathbf{I} - e^{\mathbf{A}(\theta_1+\theta_2)})^{-1}\left[-e^{\mathbf{A}\theta_1}\mathbf{A}^{-1}(e^{\mathbf{A}\theta_2} - \mathbf{I}) + \mathbf{A}^{-1}(e^{\mathbf{A}\theta_1} - \mathbf{I})\right]\mathbf{B}.$$

Considering that $\mathbf{A}e^{\mathbf{A}t}\mathbf{A}^{-1} = e^{\mathbf{A}t}$, we obtain the following formula,

$$\eta_p = (\mathbf{I} - e^{\mathbf{A}(\theta_1+\theta_2)})^{-1}\mathbf{A}^{-1}\left[-e^{\mathbf{A}\theta_1}(e^{\mathbf{A}\theta_2} - \mathbf{I}) + e^{\mathbf{A}\theta_1} - \mathbf{I}\right]\mathbf{B}$$
$$= (\mathbf{I} - e^{\mathbf{A}(\theta_1+\theta_2)})^{-1}\mathbf{A}^{-1}\left[2e^{\mathbf{A}\theta_1} - e^{\mathbf{A}(\theta_1+\theta_2)} - \mathbf{I}\right]\mathbf{B}.$$

We similarly substitute (10.2) for η_p in (10.3),

$$\rho_p = e^{\mathbf{A}\theta_2}\left[e^{\mathbf{A}\theta_1}\rho_p + \mathbf{A}^{-1}(e^{\mathbf{A}\theta_1} - \mathbf{I})\mathbf{B}\right] - \mathbf{A}^{-1}(e^{\mathbf{A}\theta_2} - \mathbf{I})\mathbf{B}$$
$$= e^{\mathbf{A}(\theta_1+\theta_2)}\rho_p + e^{\mathbf{A}\theta_2}\mathbf{A}^{-1}(e^{\mathbf{A}\theta_1} - \mathbf{I})\mathbf{B} - \mathbf{A}^{-1}(e^{\mathbf{A}\theta_2} - \mathbf{I})\mathbf{B}$$

or

$$(\mathbf{I} - e^{\mathbf{A}(\theta_1+\theta_2)})\rho_p = e^{\mathbf{A}\theta_2}\mathbf{A}^{-1}(e^{\mathbf{A}\theta_1} - \mathbf{I})\mathbf{B} - \mathbf{A}^{-1}(e^{\mathbf{A}\theta_2} - \mathbf{I})\mathbf{B},$$

which gives

$$\rho_p = (\mathbf{I} - e^{\mathbf{A}(\theta_1+\theta_2)})^{-1}\left[e^{\mathbf{A}\theta_2}\mathbf{A}^{-1}(e^{\mathbf{A}\theta_1} - \mathbf{I}) - \mathbf{A}^{-1}(e^{\mathbf{A}\theta_2} - \mathbf{I})\right]\mathbf{B}$$
$$= (\mathbf{I} - e^{\mathbf{A}(\theta_1+\theta_2)})^{-1}\left[e^{\mathbf{A}\theta_2}\mathbf{A}^{-1}e^{\mathbf{A}\theta_1} - e^{\mathbf{A}\theta_2}\mathbf{A}^{-1} - \mathbf{A}^{-1}e^{\mathbf{A}\theta_2} + \mathbf{A}^{-1}\right]\mathbf{B}$$
$$= (\mathbf{I} - e^{\mathbf{A}(\theta_1+\theta_2)})^{-1}\left[\mathbf{A}^{-1}\mathbf{A}e^{\mathbf{A}\theta_2}\mathbf{A}^{-1}e^{\mathbf{A}\theta_1}\right.$$
$$\left. - \mathbf{A}^{-1}\mathbf{A}e^{\mathbf{A}\theta_2}\mathbf{A}^{-1} - \mathbf{A}^{-1}e^{\mathbf{A}\theta_2} + \mathbf{A}^{-1}\right]\mathbf{B}$$
$$= (\mathbf{I} - e^{\mathbf{A}(\theta_1+\theta_2)})^{-1}\mathbf{A}^{-1}\left[e^{\mathbf{A}\theta_2}e^{\mathbf{A}\theta_1} - e^{\mathbf{A}\theta_2} - e^{\mathbf{A}\theta_2} + \mathbf{I}\right]\mathbf{B}$$
$$= (\mathbf{I} - e^{\mathbf{A}(\theta_1+\theta_2)})^{-1}\mathbf{A}^{-1}\left[e^{\mathbf{A}(\theta_1+\theta_2)} - 2e^{\mathbf{A}\theta_2} + \mathbf{I}\right]\mathbf{B}.$$

Considering that $\theta_1 + \theta_2 = T$, the solution of (10.2), (10.3) results in:

$$\eta_p = (\mathbf{I} - e^{\mathbf{A}T})^{-1}\mathbf{A}^{-1}[2e^{\mathbf{A}\theta_1} - e^{\mathbf{A}T} - \mathbf{I}]\mathbf{B}, \tag{10.4}$$

$$\rho_p = (\mathbf{I} - e^{\mathbf{A}T})^{-1}\mathbf{A}^{-1}[e^{\mathbf{A}T} - 2e^{\mathbf{A}\theta_2} + \mathbf{I}]\mathbf{B}. \tag{10.5}$$

Consider now the symmetric motion as a limit at $\theta_1; \theta_2 \to \theta = T/2$,

$$\lim_{\theta_1;\theta_2 \to \theta = T/2} \rho_p = (\mathbf{I} - e^{2\mathbf{A}\theta})^{-1}\mathbf{A}^{-1}[e^{2\mathbf{A}\theta} - 2e^{\mathbf{A}\theta} + \mathbf{I}]\mathbf{B}$$
$$= (\mathbf{I} - e^{2\mathbf{A}\theta})^{-1}\mathbf{A}^{-1}(\mathbf{I} - e^{\mathbf{A}\theta})^2\mathbf{B}$$
$$= (\mathbf{I} - e^{2\mathbf{A}\theta})^{-1}(\mathbf{I} - e^{\mathbf{A}\theta})^2\mathbf{A}^{-1}\mathbf{B}$$
$$= (\mathbf{I} + e^{\mathbf{A}\theta})^{-1}(\mathbf{I} - e^{\mathbf{A}\theta})\mathbf{A}^{-1}\mathbf{B}.$$
$$\tag{10.6}$$

10.1 The LPRS derivation for a non-integrating linear part

The imaginary part of the LPRS can be obtained from (10.6) in accordance with its definition as follows:

$$\mathrm{Im} J(\omega) = \frac{\pi}{4}\mathbf{C} \lim_{\theta_1;\theta_2 \to \theta=T/2} \rho_p = \frac{\pi}{4}\mathbf{C}(\mathbf{I}+e^{\mathbf{A}\pi/\omega})^{-1}(\mathbf{I}-e^{\mathbf{A}\pi/\omega})\mathbf{A}^{-1}\mathbf{B}. \quad (10.7)$$

For deriving the expression of the real part of the LPRS, consider the periodic solution (10.4) and (10.5) as a result of the feedback action,

$$\begin{cases} f_0 - y(0) = b \\ f_0 - y(\theta_1) = -b. \end{cases} \quad (10.8)$$

Having solved the set of equations (10.8) for f_0, we obtain

$$f_0 = \frac{y(0)+y(\theta_1)}{2}.$$

Hence, the constant term of the error signal $\sigma(t)$ is

$$\sigma_0 = f_0 - y_0 = \frac{y(0)+y(\theta_1)}{2} - y_0.$$

The real part of the LPRS definition formula can be transformed into

$$\mathrm{Re}\, J(\omega) = -0.5 \lim_{\gamma \to \frac{1}{2}} \frac{0.5\,[y(0)+y(\theta_1)] - y_0}{u_0}, \quad (10.9)$$

where $\gamma = \frac{\theta_1}{\theta_1+\theta_2} = \frac{\theta_1}{T}$.

Then $\theta_1 = \gamma T$, $\theta_2 = (1-\gamma)T$, $u_0 = 2\gamma-1$, and (10.9) can be rewritten as

$$\mathrm{Re}\, J(\omega) = -0.5 \lim_{\gamma \to \frac{1}{2}} \frac{0.5\mathbf{C}\,[\rho_p + \eta_p] - y_0}{2\gamma-1} \quad (10.10)$$

where ρ_p and η_p are given by (10.5) and (10.4), respectively. We find some auxiliary limits that will be instrumental:

$$\lim_{u_0 \to 0 (\theta_1+\theta_2-T-const)} \frac{e^{\mathbf{A}\theta_1}-e^{\mathbf{A}\theta_2}}{u_0} = \lim_{\gamma \to \frac{1}{2}(\theta_1+\theta_2-T-const)} \frac{e^{\mathbf{A}\theta_1}-e^{\mathbf{A}\theta_2}}{u_0}$$

$$= \lim_{\gamma \to \frac{1}{2}(\theta_1+\theta_2=T=const)} \frac{e^{\mathbf{A}\theta_1}-e^{\mathbf{A}\theta_2}}{2\gamma-1} = \lim_{\gamma \to \frac{1}{2}} \frac{e^{\mathbf{A}\gamma T}-e^{\mathbf{A}(1-\gamma)T}}{2\gamma-1}$$

$$= \lim_{\gamma \to \frac{1}{2}} \frac{e^{\mathbf{A}\gamma T}-e^{-\mathbf{A}\gamma T}e^{\mathbf{A}T}}{2\gamma-1}.$$

Taking derivatives of the numerator and the denominator, we find the limit:

$$\lim_{\gamma \to \frac{1}{2}} \frac{e^{\mathbf{A}\gamma T}-e^{-\mathbf{A}\gamma T}e^{\mathbf{A}T}}{2\gamma-1} = \lim_{\gamma \to \frac{1}{2}} \frac{\mathbf{A}Te^{\mathbf{A}\gamma T}+\mathbf{A}Te^{-\mathbf{A}\gamma T}e^{\mathbf{A}T}}{2}$$

$$= \frac{\mathbf{A}Te^{\mathbf{A}T/2}+\mathbf{A}Te^{-\mathbf{A}T/2}e^{\mathbf{A}T}}{2} = \mathbf{A}Te^{\mathbf{A}T/2}. \quad (10.11)$$

Now using formula (10.10), we find the following limit,

$$\lim_{u_0 \to 0 (\theta_1+\theta_2=T=const)} \frac{\rho_p+\eta_p}{u_0} = \lim_{\gamma \to \frac{1}{2}} \frac{\rho_p+\eta_p}{2\gamma-1}$$
$$= \lim_{\gamma \to \frac{1}{2}} \frac{2(\mathbf{I}-e^{\mathbf{A}T})^{-1}\mathbf{A}^{-1}[e^{\mathbf{A}\theta_1}-e^{\mathbf{A}\theta_2}]\mathbf{B}}{2\gamma-1}$$
$$= 2(\mathbf{I}-e^{\mathbf{A}T})^{-1}\mathbf{A}^{-1} \lim_{\gamma \to \frac{1}{2}} \frac{e^{\mathbf{A}\theta_1}-e^{\mathbf{A}\theta_2}}{2\gamma-1}\mathbf{B} \qquad (10.12)$$
$$= 2(\mathbf{I}-e^{\mathbf{A}T})^{-1}\mathbf{A}^{-1}\mathbf{A}Te^{\mathbf{A}T/2}\mathbf{B}$$
$$= 2T(\mathbf{I}-e^{\mathbf{A}T})^{-1}e^{\mathbf{A}T/2}\mathbf{B}.$$

To calculate the limit $\lim\limits_{u_0 \to 0} \frac{y_0}{u_0}$, consider the equations for the constant terms of the variables (averaged variables), which are obtained from the original equations of the plant by equating the derivatives to zero,

$$\begin{cases} 0 = \mathbf{A}\mathbf{x}_0 + \mathbf{B}u_0 \\ y_0 = \mathbf{C}\mathbf{x}_0. \end{cases}$$

From those equations, we obtain $\mathbf{x}_0 = -\mathbf{A}^{-1}\mathbf{B}u_0$ and $y_0 = -\mathbf{C}\mathbf{A}^{-1}\mathbf{B}u_0$. Therefore,

$$\lim_{u_0 \to 0} \frac{y_0}{u_0} = -\mathbf{C}\mathbf{A}^{-1}\mathbf{B}, \qquad (10.13)$$

which is essentially the steady-state gain of the plant. The real part of the LPRS is obtained by substituting (10.12) and (10.13) for the respective limits in (10.10),

$$\mathrm{Re}\, J(\omega) = -0.5 \lim_{u_0 \to 0} \frac{0.5\mathbf{C}\left[\rho_p+\eta_p\right]-y_0}{u_0}$$
$$= -0.25\mathbf{C} \lim_{u_0 \to 0} \frac{\rho_p+\eta_p}{u_0} + 0.5 \lim_{u_0 \to 0} \frac{y_0}{u_0} \qquad (10.14)$$
$$= -0.5T\mathbf{C}(\mathbf{I}-e^{\mathbf{A}T})^{-1}e^{\mathbf{A}T/2}\mathbf{B} - 0.5\mathbf{C}\mathbf{A}^{-1}\mathbf{B}.$$

The real part of the LPRS was derived under the assumption that the limits at $u_0 \to 0$ and at $\gamma \to \frac{1}{2}$ are equal. This is true only if the period T does not change. Therefore, we need to prove that this is the case. The frequency of the oscillations (or the period) is defined by the following switching condition obtained from (10.8),

$$y(0) - y(\theta_1) = -2b.$$

We find the derivative $\frac{\partial(y(0)-y(\theta_1))}{\partial u_0} - \frac{1}{2}\frac{\partial(y(0)-y(\theta_1))}{\partial \gamma}$ at the point $\gamma = \frac{1}{2}$,

$$y(0) - y(\theta_1) = \mathbf{C}(\mathbf{I}-e^{\mathbf{A}T})^{-1}\mathbf{A}^{-1}$$
$$\times [e^{\mathbf{A}T} - 2e^{\mathbf{A}\theta_2} + \mathbf{I} - 2e^{\mathbf{A}\theta_1} + e^{\mathbf{A}T} + \mathbf{I}]\mathbf{B}$$
$$= 2\mathbf{C}(\mathbf{I}-e^{\mathbf{A}T})^{-1}\mathbf{A}^{-1}\left[e^{\mathbf{A}T} + \mathbf{I} - \left(e^{\mathbf{A}\theta_1} + e^{\mathbf{A}\theta_2}\right)\right]\mathbf{B}$$
$$= 2\mathbf{C}(\mathbf{I}-e^{\mathbf{A}T})^{-1}\mathbf{A}^{-1}\left[e^{\mathbf{A}T} + \mathbf{I} - \left(e^{\mathbf{A}\gamma T} + e^{\mathbf{A}(1-\gamma)T}\right)\right]\mathbf{B}.$$

10.2 Orbital stability of a system with a non-integrating linear part

We take the derivative with respect to γ of the last formula:

$$\frac{\partial (y(0) - y(\theta_1))}{\partial \gamma} = -2\mathbf{C}(\mathbf{I} - e^{\mathbf{A}T})^{-1}\mathbf{A}^{-1}\left[\mathbf{A}Te^{\mathbf{A}\gamma T} - \mathbf{A}Te^{\mathbf{A}(1-\gamma)T}\right]\mathbf{B}.$$

$$= -2\mathbf{C}(\mathbf{I} - e^{\mathbf{A}T})^{-1}T\left[e^{\mathbf{A}\gamma T} - e^{\mathbf{A}(1-\gamma)T}\right]\mathbf{B}$$

The derivative $\frac{\partial (y(0) - y(\theta_1))}{\partial \gamma}$ at the point $\gamma = \frac{1}{2}$ is as follows:

$$\left.\frac{\partial (y(0) - y(\theta_1))}{\partial \gamma}\right|_{\gamma = 1/2} = 0.$$

It follows from the last formula that the two limits are equivalent. It also follows from the last formula that

$$\left.\frac{d\Omega}{df_0}\right|_{f_0 = 0} = 0.$$

Finally, the state-space description–based formula of the LPRS can be obtained by combining formulas (10.7) and (10.15) as follows:

$$J(\omega) = -0.5\mathbf{C}[\mathbf{A}^{-1} + \frac{2\pi}{\omega}(\mathbf{I} - e^{\frac{2\pi}{\omega}\mathbf{A}})^{-1}e^{\frac{\pi}{\omega}\mathbf{A}}]\mathbf{B}$$
$$+ j\frac{\pi}{4}\mathbf{C}(\mathbf{I} + e^{\frac{\pi}{\omega}\mathbf{A}})^{-1}(\mathbf{I} - e^{\frac{\pi}{\omega}\mathbf{A}})\mathbf{A}^{-1}\mathbf{B}. \tag{10.15}$$

10.2 Orbital stability of a system with a non-integrating linear part

Assume that the periodic solution is given by formulas (10.4) and (10.5), and at the initial time $t = 0$ the state vector has a deviation from the value in a periodic motion, which is given as $\mathbf{x}(0) = \rho = \rho_p + \delta\rho$. Then for the time interval $t \in [0; t^*]$, where t^* is the time of the switch of the relay from "+1" to "-1," the state vector (when the control is $u = 1$) is given as follows:

$$\mathbf{x}(t) = e^{\mathbf{A}t}(\rho_p + \delta\rho) + \mathbf{A}^{-1}(e^{\mathbf{A}t} - \mathbf{I})\mathbf{B}$$
$$= e^{\mathbf{A}t}\rho_p + \mathbf{A}^{-1}(e^{\mathbf{A}t} - \mathbf{I})\mathbf{B} + e^{\mathbf{A}t}\delta\rho, \tag{10.16}$$

where the first two addends give the unperturbed motion, and the third addend gives the motion due to the initial perturbation $\delta\rho$. We denote

$$\mathbf{x}(t^*) = \eta = \eta_p + \delta\eta. \tag{10.17}$$

We note that η_p is determined not at time $t = t^*$ but at time $t = \theta_1$.

The main task of our analysis is to find the mapping $\delta\rho \to \delta\eta$. Therefore, it follows from (10.17) that

$$\delta\eta = \mathbf{x}(t^*) - \eta_p. \tag{10.18}$$

Assuming that all perturbations are small and times t^* and θ_1 are close, we evaluate $\mathbf{x}(t^*)$ by a linear approximation of $\mathbf{x}(t)$ at the point $t = \theta_1$,

$$\mathbf{x}(t^*) - \mathbf{x}(\theta_1) = \dot{\mathbf{x}}(\theta_1-)(t^* - \theta_1),$$

where $\dot{\mathbf{x}}(\theta_1-)$ is the value of the derivative at the time immediately preceding the time $t = \theta_1$. We express $\mathbf{x}(t^*)$ from the last equation as follows,

$$\mathbf{x}(t^*) = \mathbf{x}(\theta_1) + \dot{\mathbf{x}}(\theta_1-)(t^* - \theta_1). \tag{10.19}$$

Now we evaluate $\mathbf{x}(\theta_1)$ using (10.16).

$$\mathbf{x}(\theta_1) = \eta_p + e^{\mathbf{A}\theta_1}\delta\rho. \tag{10.20}$$

We substitute (10.19) and (10.20) in (10.18),

$$\begin{aligned}\delta\eta &= \mathbf{x}(\theta_1) + \dot{\mathbf{x}}(\theta_1-)(t^* - \theta_1) - \eta_p \\ &= \eta_p + e^{\mathbf{A}\theta_1}\delta\rho + \dot{\mathbf{x}}(\theta_1-)(t^* - \theta_1) - \eta_p \\ &= \mathbf{v}(\theta_1-)(t^* - \theta_1) + e^{\mathbf{A}\theta_1}\delta\rho,\end{aligned} \tag{10.21}$$

where $\mathbf{v}(t) = \dot{\mathbf{x}}(t)$. We evaluate $(t^* - \theta_1)$. The system output $y(t)$ crosses the level $f_0 + b$ at $t = t^*$, which results in the switch. Using linear approximation, we can write for the perturbation of the output:

$$(t^* - \theta_1)\dot{y}(\theta_1-) = -\delta y(\theta_1) = -\mathbf{C}\delta\mathbf{x}(\theta_1-) = -\mathbf{C}e^{\mathbf{A}\theta_1}\delta\rho. \tag{10.22}$$

From (10.22), we obtain

$$t^* - \theta_1 = -\frac{\delta y(\theta_1)}{\dot{y}(\theta_1-)} = -\frac{\mathbf{C}e^{\mathbf{A}\theta_1}}{\dot{y}(\theta_1-)}\delta\rho.$$

Now we substitute the expression for $t^* - \theta_1$ in (10.22),

$$\begin{aligned}\delta\eta &= -\mathbf{v}(\theta_1-)\frac{\mathbf{C}e^{\mathbf{A}\theta_1}}{\dot{y}(\theta_1-)}\delta\rho + e^{\mathbf{A}\theta_1}\delta\rho = e^{\mathbf{A}\theta_1}\delta\rho - \mathbf{v}(\theta_1-)\frac{\mathbf{C}e^{\mathbf{A}\theta_1}}{\mathbf{C}\mathbf{v}(\theta_1-)}\delta\rho \\ &= \left[e^{\mathbf{A}\theta_1} - \frac{\mathbf{v}(\theta_1-)\mathbf{C}}{\mathbf{C}\mathbf{v}(\theta_1-)}e^{\mathbf{A}\theta_1}\right]\delta\rho = \left[\mathbf{I} - \frac{\mathbf{v}(\theta_1-)\mathbf{C}}{\mathbf{C}\mathbf{v}(\theta_1-)}\right]e^{\mathbf{A}\theta_1}\delta\rho.\end{aligned} \tag{10.23}$$

Denote

$$\Phi_1 = \left[\mathbf{I} - \frac{\mathbf{v}(\theta_1-)\mathbf{C}}{\mathbf{C}\mathbf{v}(\theta_1-)}\right]e^{\mathbf{A}\theta_1}, \tag{10.24}$$

which is the Jacobian matrix of the mapping $\delta\rho \to \delta\eta$. The velocity vector $\mathbf{v}(\theta_1-)$ in (10.24) is determined by the following formula:

$$\mathbf{v}(\theta_1 -) = \mathbf{A}\mathbf{x}(\theta_1) + \mathbf{B}$$

$$= \mathbf{A}\left(\mathbf{I} - e^{\mathbf{A}T}\right)^{-1} \mathbf{A}^{-1} \left[2e^{\mathbf{A}\theta_1} - e^{\mathbf{A}T} - \mathbf{I}\right]\mathbf{B} + \mathbf{B}$$

$$= \left(\mathbf{I} - e^{\mathbf{A}T}\right)^{-1} \left[2e^{\mathbf{A}\theta_1} - e^{\mathbf{A}T} - \mathbf{I}\right]\mathbf{B} + \mathbf{B}$$

$$= \left\{\left(\mathbf{I} - e^{\mathbf{A}T}\right)^{-1} \left[2e^{\mathbf{A}\theta_1} - e^{\mathbf{A}T} - \mathbf{I}\right] + \mathbf{I}\right\}\mathbf{B} \quad (10.25)$$

$$= \left\{\left(\mathbf{I} - e^{\mathbf{A}T}\right)^{-1} \left[2e^{\mathbf{A}\theta_1} - e^{\mathbf{A}T} - \mathbf{I} + (\mathbf{I} - e^{\mathbf{A}T})\right]\right\}\mathbf{B}$$

$$= \left\{\left(\mathbf{I} - e^{\mathbf{A}T}\right)^{-1} \left[2e^{\mathbf{A}\theta_1} - 2e^{\mathbf{A}T}\right]\right\}\mathbf{B}$$

$$= 2\left(\mathbf{I} - e^{\mathbf{A}T}\right)^{-1} \left(e^{\mathbf{A}\theta_1} - e^{\mathbf{A}T}\right)\mathbf{B}.$$

To be able to assess local orbital stability of the system, we need to find the Jacobian matrix of the Poincaré return map, which arises from the chain-rule application of two mappings: $\delta\rho \to \delta\eta$ and $\delta\eta \to \delta\rho$. Reasoning along the same lines, we find the Jacobian matrix of the mapping $\delta\eta \to \delta\rho$,

$$\Phi_2 = \left[\mathbf{I} - \frac{\mathbf{v}(\theta_2-)\mathbf{C}}{\mathbf{C}\mathbf{v}(\theta_2-)}\right] e^{\mathbf{A}\theta_2}, \quad (10.26)$$

where

$$\mathbf{v}(\theta_2-) = 2\left(\mathbf{I} - e^{\mathbf{A}T}\right)^{-1} \left(e^{\mathbf{A}T} - e^{\mathbf{A}\theta_2}\right)\mathbf{B}. \quad (10.27)$$

The local orbital stability of the system is determined by the eigenvalues of the matrix $\Phi = \Phi_2\Phi_1$. If all eigenvalues of Φ have magnitudes less than *one*, then the system is orbitally asymptotically stable.

For the symmetric motion, it is sufficient to check only half of the period of motion. If, therefore, all eigenvalues of the matrix

$$\Phi_0 = \left[\mathbf{I} - \frac{\mathbf{v}(\frac{T}{2}-)\mathbf{C}}{\mathbf{C}\mathbf{v}(\frac{T}{2}-)}\right] e^{\mathbf{A}\frac{T}{2}}, \quad (10.28)$$

where

$$\mathbf{v}(\frac{T}{2}-) = 2\left(\mathbf{I} - e^{\mathbf{A}T}\right)^{-1} \left(e^{\mathbf{A}\frac{T}{2}} - e^{\mathbf{A}T}\right)\mathbf{B} = 2\left(\mathbf{I} + e^{\mathbf{A}T/2}\right)^{-1} e^{\mathbf{A}T/2}\mathbf{B},$$

have magnitudes less than *one*, then the system is locally orbitally asymptotically stable.

10.3 The LPRS derivation for an integrating linear part

We derive the formula for $J(\omega)$ in the case of an integrating linear part (*type 1 relay servo system*). The state-space description of the system has the following form:

$$\dot{\mathbf{x}} = \mathbf{A}\mathbf{x} + \mathbf{B}u$$
$$\dot{y} = \mathbf{C}\mathbf{x} - f_0 \qquad (10.29)$$

$$u = \begin{cases} +1 \text{ if } \sigma = -y \geq b & \text{or} \quad \sigma > -b, \ u(t-) = 1 \\ -1 \text{ if } \sigma = -y \leq -b & \text{or} \quad \sigma < b, \ u(t-) = -1 \end{cases}$$

where $\mathbf{A} \in R^{(n-1) \times (n-1)}$, $\mathbf{B} \in R^{(n-1) \times 1}$, $\mathbf{C} \in R^{1 \times (n-1)}$, f_0 is a constant input to the system, $\sigma(t)$ is the error signal, $2b$ is the hysteresis of the relay function, and $\mathbf{A}$ is nonsingular.

The periodic solution for $\mathbf{x}(t)$ before the integrator is given above (formulas (10.4) and (10.5)). The periodic output $y(t)$ can be obtained by integrating the second equation of (10.29) from the initial states determined by formulas (10.4) and (10.5). As a result, for the control $u = 1$, the system output can be written as follows (denote it as $y_1(t)$),

$$y_1(t) = y(0) + \int_0^t (\mathbf{C}\mathbf{x}(\tau) - f_0) \, d\tau$$
$$= y(0) + \int_0^t \left(\mathbf{C}(e^{\mathbf{A}\tau}\rho_p + \mathbf{A}^{-1}(e^{\mathbf{A}\tau} - \mathbf{I})\mathbf{B}) - f_0 \right) d\tau$$
$$= y(0) + \mathbf{C}\left[\mathbf{A}^{-1}e^{\mathbf{A}t}\rho_p \big|_0^t + \mathbf{A}^{-2}e^{\mathbf{A}t}\mathbf{B} \big|_0^t - \mathbf{A}^{-1}\mathbf{B}t \big|_0^t \right] - f_0 t \big|_0^t$$
$$= y(0) + \mathbf{C}\left[\mathbf{A}^{-1}e^{\mathbf{A}t}\rho_p - \mathbf{A}^{-1}\rho_p + \mathbf{A}^{-2}e^{\mathbf{A}t}\mathbf{B} - \mathbf{A}^{-2}\mathbf{B} - \mathbf{A}^{-1}\mathbf{B}t \right] - f_0 t$$
$$= y(0) - \mathbf{C}\mathbf{A}^{-1}\mathbf{B}t - f_0 t + \mathbf{C}\mathbf{A}^{-1}[(e^{\mathbf{A}t} - \mathbf{I})\rho_p + \mathbf{A}^{-1}(e^{\mathbf{A}t} - \mathbf{I})\mathbf{B}]. \qquad (10.30)$$

From (10.30), we find the value of the output at $t = \theta_1$:

$$y_1(\theta_1) = y(0) - (\mathbf{C}\mathbf{A}^{-1}\mathbf{B} + f_0)\theta_1 + \mathbf{C}\mathbf{A}^{-1}[(e^{\mathbf{A}\theta_1} - \mathbf{I})\rho_p + \mathbf{A}^{-1}(e^{\mathbf{A}\theta_1} - \mathbf{I})\mathbf{B}]. \qquad (10.31)$$

We write the system output for the control $u = -1$ (denote it $y_2(t')$, where $t' = t - \theta_1$ is a shifted time):

$$y_2(t') = y_1(\theta_1) + \int_0^{t'} (\mathbf{C}\mathbf{x}(\tau) - f_0) \, d\tau$$
$$= y_1(\theta_1) + \int_0^{t'} \left(\mathbf{C}(e^{\mathbf{A}\tau}\eta_p - \mathbf{A}^{-1}(e^{\mathbf{A}\tau} - \mathbf{I})\mathbf{B}) - f_0 \right) d\tau$$
$$= y_1(\theta_1) + \mathbf{C}\left[\mathbf{A}^{-1}e^{\mathbf{A}t}\eta_p \big|_0^{t'} - \mathbf{A}^{-2}e^{\mathbf{A}t}\mathbf{B} \big|_0^{t'} + \mathbf{A}^{-1}\mathbf{B}t \big|_0^{t'} \right] - f_0 t \big|_0^{t'}$$
$$= y_1(\theta_1) + \mathbf{C}\left[\mathbf{A}^{-1}e^{\mathbf{A}t'}\eta_p - \mathbf{A}^{-1}\eta_p - \mathbf{A}^{-2}e^{\mathbf{A}t'}\mathbf{B} + \mathbf{A}^{-2}\mathbf{B} + \mathbf{A}^{-1}\mathbf{B}t' \right] - f_0 t'$$
$$= y_1(\theta_1) + \mathbf{C}\mathbf{A}^{-1}\mathbf{B}t' - f_0 t' + \mathbf{C}\mathbf{A}^{-1}[(e^{\mathbf{A}t'} - \mathbf{I})\eta_p - \mathbf{A}^{-1}(e^{\mathbf{A}t'} - \mathbf{I})\mathbf{B}]. \qquad (10.32)$$

From (10.32), we find the value of the output at $t' = \theta_2$:

$$y_2(\theta_2) = y_1(\theta_1) + (\mathbf{C}\mathbf{A}^{-1}\mathbf{B} - f_0)\theta_2 + \mathbf{C}\mathbf{A}^{-1}[(e^{\mathbf{A}\theta_2} - \mathbf{I})\eta_p - \mathbf{A}^{-1}(e^{\mathbf{A}\theta_2} - \mathbf{I})\mathbf{B}]. \qquad (10.33)$$

Because $y_1(t) = y(t)$ and $y_2(t') = y(t + \theta_1)$, the last formula gives $y_2(\theta_2) = y(\theta_1 + \theta_2) = y(T) = y(0)$. We rewrite (10.31), (10.33) as follows:

10.3 The LPRS derivation for an integrating linear part

$$y(\theta_1) - y(0) + (\mathbf{CA}^{-1}\mathbf{B} + f_0)\theta_1 = \mathbf{CA}^{-1}[(e^{\mathbf{A}\theta_1} - \mathbf{I})\rho_p + \mathbf{A}^{-1}(e^{\mathbf{A}\theta_1} - \mathbf{I})\mathbf{B}] \tag{10.34}$$

$$y(0) - y(\theta_1) - (\mathbf{CA}^{-1}\mathbf{B} - f_0)\theta_2 = \mathbf{CA}^{-1}[(e^{\mathbf{A}\theta_2} - \mathbf{I})\eta_p - \mathbf{A}^{-1}(e^{\mathbf{A}\theta_2} - \mathbf{I})\mathbf{B}]. \tag{10.35}$$

The constant input f_0 must be equal to the constant term of the output of the plant (excluding the integrator), so that the constant term (averaged value) of the input to the integrator is zero,

$$f_0 = -\mathbf{CA}^{-1}\mathbf{B}(2\gamma - 1).$$

We assess the components in (10.34), (10.35) that contain f_0. From (10.35) we have:

$$(\mathbf{CA}^{-1}\mathbf{B} + f_0)\theta_1 = (\mathbf{CA}^{-1}\mathbf{B} - \mathbf{CA}^{-1}\mathbf{B}(2\gamma - 1))\gamma T$$
$$= \mathbf{CA}^{-1}\mathbf{B}(1 - 2\gamma + 1)\gamma T = 2\mathbf{CA}^{-1}\mathbf{B}(1 - \gamma)\gamma T.$$

From (10.34), we have the following:

$$(\mathbf{CA}^{-1}\mathbf{B} - f_0)\theta_2 = (\mathbf{CA}^{-1}\mathbf{B} + \mathbf{CA}^{-1}\mathbf{B}(2\gamma - 1))(1 - \gamma)T$$
$$= \mathbf{CA}^{-1}\mathbf{B}(1 + 2\gamma - 1)(1 - \gamma)T = 2\mathbf{CA}^{-1}\mathbf{B}(1 - \gamma)\gamma T.$$

One can see that both components are equal. We rewrite (10.34), (10.35) accounting for the last formulas:

$$y(\theta_1) - y(0) + 2\gamma(1-\gamma)T\mathbf{CA}^{-1}\mathbf{B} = \mathbf{CA}^{-1}[(e^{\mathbf{A}\theta_1} - \mathbf{I})\rho_p + \mathbf{A}^{-1}(e^{\mathbf{A}\theta_1} - \mathbf{I})\mathbf{B}] \tag{10.36}$$

$$y(0) - y(\theta_1) - 2\gamma(1-\gamma)T\mathbf{CA}^{-1}\mathbf{B} = \mathbf{CA}^{-1}[(e^{\mathbf{A}\theta_2} - \mathbf{I})\eta_p - \mathbf{A}^{-1}(e^{\mathbf{A}\theta_2} - \mathbf{I})\mathbf{B}]. \tag{10.37}$$

We take the sum of the left-hand sides and the right-hand sides of (10.36) and (10.37),

$$0 = \mathbf{CA}^{-1}[(e^{\mathbf{A}\theta_1} - \mathbf{I})\rho_p + \mathbf{A}^{-1}(e^{\mathbf{A}\theta_1} - \mathbf{I})\mathbf{B}$$
$$+ (e^{\mathbf{A}\theta_2} - \mathbf{I})\eta_p - \mathbf{A}^{-1}(e^{\mathbf{A}\theta_2} - \mathbf{I})\mathbf{B}]$$
$$= \mathbf{C}[\mathbf{A}^{-1}((e^{\mathbf{A}\theta_1} - \mathbf{I})\rho_p + (e^{\mathbf{A}\theta_2} - \mathbf{I})\eta_p)$$
$$+ \mathbf{A}^{-2}((e^{\mathbf{A}\theta_1} - \mathbf{I}) - (e^{\mathbf{A}\theta_2} - \mathbf{I}))\mathbf{B}]$$
$$= \mathbf{C}[\mathbf{A}^{-2}(\mathbf{I} - e^{\mathbf{A}T})^{-1}[(e^{\mathbf{A}\theta_1} - \mathbf{I})(e^{\mathbf{A}T^*} - 2e^{\mathbf{A}\theta_2} + \mathbf{I})$$
$$+ (e^{\mathbf{A}\theta_2} - \mathbf{I})(2e^{\mathbf{A}\theta_1} - e^{\mathbf{A}T} - \mathbf{I})]\mathbf{B} + \mathbf{A}^{-2}(e^{\mathbf{A}\theta_1} - e^{\mathbf{A}\theta_2})\mathbf{B}]$$
$$= \mathbf{C}[\mathbf{A}^{-2}(\mathbf{I} - e^{\mathbf{A}T})^{-1}[e^{\mathbf{A}\theta_1}e^{\mathbf{A}T} - 2e^{\mathbf{A}T} + e^{\mathbf{A}\theta_1} - e^{\mathbf{A}T} + 2e^{\mathbf{A}\theta_2} - \mathbf{I}$$

$$+2e^{\mathbf{A}T} - e^{\mathbf{A}\theta_2}e^{\mathbf{A}T} - e^{\mathbf{A}\theta_2} - 2e^{\mathbf{A}\theta_1} + e^{\mathbf{A}T} + \mathbf{I}]\mathbf{B} + \mathbf{A}^{-2}(e^{\mathbf{A}\theta_1} - e^{\mathbf{A}\theta_2})\mathbf{B}]$$
$$= \mathbf{C}[\mathbf{A}^{-2}(\mathbf{I} - e^{\mathbf{A}T})^{-1}[-e^{\mathbf{A}\theta_1} + e^{\mathbf{A}\theta_2} + e^{\mathbf{A}\theta_1}e^{\mathbf{A}T} - e^{\mathbf{A}\theta_2}e^{\mathbf{A}T}]\mathbf{B}$$
$$+\mathbf{A}^{-2}(e^{\mathbf{A}\theta_1} - e^{\mathbf{A}\theta_2})\mathbf{B}]$$
$$= \mathbf{C}[\mathbf{A}^{-2}(\mathbf{I} - e^{\mathbf{A}T})^{-1}(e^{\mathbf{A}T} - \mathbf{I})(e^{\mathbf{A}\theta_1} - e^{\mathbf{A}\theta_2})\mathbf{B} + \mathbf{A}^{-2}(e^{\mathbf{A}\theta_1} - e^{\mathbf{A}\theta_2})\mathbf{B}]$$
$$= \mathbf{C}[-\mathbf{A}^{-2}(e^{\mathbf{A}\theta_1} - e^{\mathbf{A}\theta_2})\mathbf{B} + \mathbf{A}^{-2}(e^{\mathbf{A}\theta_1} - e^{\mathbf{A}\theta_2})\mathbf{B}] = 0. \quad (10.38)$$

Therefore, equations (10.34) and (10.35) are equivalent and do not comprise a system with a unique solution. In fact, they allow for an infinite number of solutions. To define a unique solution, we consider the condition of the relay switching,
$$y(\theta_1) = -y(0). \quad (10.39)$$

Assuming (10.39), we take the difference between the left-hand sides and the right-hand sides of (10.35) and (10.34),

$$4y(0) - 4\gamma(1-\gamma)T\mathbf{C}\mathbf{A}^{-1}\mathbf{B} = \mathbf{C}\{A^{-1}[(e^{\mathbf{A}\theta_2} - \mathbf{I})\eta_p - (e^{\mathbf{A}\theta_1} - \mathbf{I})\rho_p]$$
$$- \mathbf{A}^{-2}[(e^{\mathbf{A}\theta_2} - \mathbf{I}) + (e^{\mathbf{A}\theta_1} - \mathbf{I})]\mathbf{B}\}$$
$$= \mathbf{C}\{\mathbf{A}^{-1}[(e^{\mathbf{A}\theta_2} - \mathbf{I})(\mathbf{I} - e^{\mathbf{A}T})^{-1}\mathbf{A}^{-1}(2e^{\mathbf{A}\theta_1} - e^{\mathbf{A}T} - \mathbf{I})\mathbf{B}$$
$$-(e^{\mathbf{A}\theta_1} - \mathbf{I})(\mathbf{I} - e^{\mathbf{A}T})^{-1}\mathbf{A}^{-1}(e^{\mathbf{A}T} - 2e^{\mathbf{A}\theta_2} + \mathbf{I})\mathbf{B}]$$
$$-\mathbf{A}^{-2}(e^{\mathbf{A}\theta_1} + e^{\mathbf{A}\theta_2} - 2\mathbf{I})\mathbf{B}\}$$
$$= \mathbf{C}\{\mathbf{A}^{-2}(\mathbf{I} - e^{\mathbf{A}T})^{-1}[(e^{\mathbf{A}\theta_2} - \mathbf{I})(2e^{\mathbf{A}\theta_1} - e^{\mathbf{A}T} - \mathbf{I}) - (e^{\mathbf{A}\theta_1} - \mathbf{I})(e^{\mathbf{A}T}$$
$$-2e^{\mathbf{A}\theta_2} + \mathbf{I})]\mathbf{B} - \mathbf{A}^{-2}(e^{\mathbf{A}\theta_1} + e^{\mathbf{A}\theta_2} - 2\mathbf{I})\mathbf{B}\}$$
$$= \mathbf{C}\{\mathbf{A}^{-2}(\mathbf{I} - e^{\mathbf{A}T})^{-1}[2e^{\mathbf{A}\theta_1}e^{\mathbf{A}\theta_2} - e^{\mathbf{A}\theta_2}e^{\mathbf{A}T} - e^{\mathbf{A}\theta_2} - 2e^{\mathbf{A}\theta_1} + e^{\mathbf{A}T} + \mathbf{I}$$
$$-e^{\mathbf{A}\theta_1}e^{\mathbf{A}T} + 2e^{\mathbf{A}\theta_1}e^{\mathbf{A}\theta_2} - e^{\mathbf{A}\theta_1} + e^{\mathbf{A}T} - 2e^{\mathbf{A}\theta_2} + \mathbf{I}]\mathbf{B}$$
$$-\mathbf{A}^{-2}(e^{\mathbf{A}\theta_1} + e^{\mathbf{A}\theta_2} - 2\mathbf{I})\mathbf{B}\}$$
$$= \mathbf{C}\{\mathbf{A}^{-2}(\mathbf{I} - e^{\mathbf{A}T})^{-1}[6e^{\mathbf{A}T} - 3e^{\mathbf{A}\theta_1} - 3e^{\mathbf{A}\theta_2}$$
$$-e^{\mathbf{A}T}(e^{\mathbf{A}\theta_1} + e^{\mathbf{A}\theta_2}) + 2\mathbf{I}]\mathbf{B} - \mathbf{A}^{-2}(e^{\mathbf{A}\theta_1} + e^{\mathbf{A}\theta_2} - 2\mathbf{I})\mathbf{B}\}$$
$$= \mathbf{C}\mathbf{A}^{-2}\{(\mathbf{I} - e^{\mathbf{A}T})^{-1}[6e^{\mathbf{A}T} - 3(e^{\mathbf{A}\theta_1} + e^{\mathbf{A}\theta_2}) - e^{\mathbf{A}T}(e^{\mathbf{A}\theta_1} + e^{\mathbf{A}\theta_2}) + 2\mathbf{I}]$$
$$-(e^{\mathbf{A}\theta_1} + e^{\mathbf{A}\theta_2}) + 2\mathbf{I}\}\mathbf{B}.$$
$$(10.40)$$

We solve (10.40) for $y(0)$,
$$y(0) = \gamma(1-\gamma)T\mathbf{C}\mathbf{A}^{-1}\mathbf{B} + \tfrac{1}{4}\mathbf{C}\mathbf{A}^{-2}\{(\mathbf{I} - e^{\mathbf{A}T})^{-1}[6e^{\mathbf{A}T} - 3(e^{\mathbf{A}\theta_1} + e^{\mathbf{A}\theta_2})$$
$$-e^{\mathbf{A}T}(e^{\mathbf{A}\theta_1} + e^{\mathbf{A}\theta_2}) + 2\mathbf{I}]$$
$$-(e^{\mathbf{A}\theta_1} + e^{\mathbf{A}\theta_2}) + 2\mathbf{I}\}\mathbf{B}.$$
$$(10.41)$$

10.3 The LPRS derivation for an integrating linear part

Formula (10.41) provides an asymmetric periodic solution in the relay system. We consider the limiting case, when $\theta_1; \theta_2 \to \theta = T/2$ $(\gamma \to \frac{1}{2})$, and we find a symmetric periodic solution:

$$\lim_{\theta_1;\theta_2 \to \theta=T/2} y(0) = \tfrac{1}{4}T\mathbf{CA}^{-1}\mathbf{B} + \tfrac{1}{4}\mathbf{CA}^{-2}\{(\mathbf{I} - e^{\mathbf{A}T})^{-1}$$
$$\times [6e^{\mathbf{A}T} - 6e^{\mathbf{A}T/2} - 2e^{\mathbf{A}T}e^{\mathbf{A}T/2} + 2\mathbf{I}] - 2e^{\mathbf{A}T/2} + 2\mathbf{I}\}\mathbf{B}$$
$$= \tfrac{1}{4}T\mathbf{CA}^{-1}\mathbf{B} + \tfrac{1}{2}\mathbf{CA}^{-2}\{(\mathbf{I} - e^{\mathbf{A}T})^{-1}$$
$$\times [3e^{\mathbf{A}T} - 3e^{\mathbf{A}T/2} - e^{\mathbf{A}T}e^{\mathbf{A}T/2} + \mathbf{I}] - e^{\mathbf{A}T/2} + \mathbf{I}\}\mathbf{B}. \tag{10.42}$$

The imaginary part of the LPRS can be obtained from (10.42) in accordance with its definition as follows:

$$\mathrm{Im}J(\omega) = \tfrac{\pi}{4} \lim_{\theta_1,\theta_2 \to \theta=T/2} y(0) = \tfrac{\pi}{16}T\mathbf{CA}^{-1}\mathbf{B} + \tfrac{\pi}{8}\mathbf{CA}^{-2}\{(\mathbf{I} - e^{\mathbf{A}T})^{-1}$$
$$\times [3e^{\mathbf{A}T} - 3e^{\mathbf{A}T/2} - e^{\mathbf{A}T}e^{\mathbf{A}T/2} + \mathbf{I}] - e^{\mathbf{A}T/2} + \mathbf{I}\}\mathbf{B}. \tag{10.43}$$

To derive the expression of the real part of the LPRS, consider the periodic solution (10.41) as a result of the feedback action. The constant term y_0 of the output $y(t)$ can be determined as the sum of integrals of functions (10.30) and (10.32), divided by period T:

$$y_0 = \frac{1}{T}\left\{\int_0^{\theta_1} y_1(\tau)d\tau + \int_0^{\theta_2} y_2(\tau)d\tau\right\}. \tag{10.44}$$

Therefore,

$$y_0 = \tfrac{1}{T}\Big\{y(0)\theta_1 + \mathbf{C}\left[\mathbf{A}^{-2}\left.e^{\mathbf{A}t}\right|_0^{\theta_1}\rho - \mathbf{A}^{-1}\rho\theta_1 + \mathbf{A}^{-3}\left.e^{\mathbf{A}t}\right|_0^{\theta_1}\mathbf{B} - \mathbf{A}^{-2}\mathbf{B}\theta_1\right]$$
$$-\tfrac{1}{2}(\mathbf{CA}^{-1}\mathbf{B} - f_0)\left.t^2\right|_0^{\theta_1}$$
$$-y(0)\theta_2 + \mathbf{C}\left[\mathbf{A}^{-2}\left.e^{\mathbf{A}t}\right|_0^{\theta_2}\eta - \mathbf{A}^{-1}\eta\theta_2 - \mathbf{A}^{-3}\left.e^{\mathbf{A}t}\right|_0^{\theta_2}\mathbf{B} + \mathbf{A}^{-2}\mathbf{B}\theta_2\right]$$
$$+\tfrac{1}{2}(\mathbf{CA}^{-1}\mathbf{B} - f_0)\left.t^2\right|_0^{\theta_2}\Big\}$$
$$= \tfrac{1}{T}\Big\{y(0)(\theta_1 - \theta_2) - \tfrac{1}{2}(\mathbf{CA}^{-1}\mathbf{B} + f_0)\theta_1^2 + \tfrac{1}{2}(\mathbf{CA}^{-1}\mathbf{B} - f_0)\theta_2^2$$
$$+\mathbf{C}\left[\mathbf{A}^{-2}(e^{\mathbf{A}\theta_1} - \mathbf{I})\rho - \mathbf{A}^{-1}\rho\theta_1 + \mathbf{A}^{-3}(e^{\mathbf{A}\theta_1} - \mathbf{I})\mathbf{B} - \mathbf{A}^{-2}\mathbf{B}\theta_1\right.$$
$$\left.-\mathbf{A}^{-1}\eta\theta_2 + \mathbf{A}^{-2}(e^{\mathbf{A}\theta_2} - \mathbf{I})\eta + \mathbf{A}^{-2}\mathbf{B}\theta_2 - \mathbf{A}^{-3}(e^{\mathbf{A}\theta_2} - \mathbf{I})\mathbf{B}\right]\Big\}$$
$$= \tfrac{1}{T}\Big\{y(0)(\theta_1 - \theta_2) + \mathbf{CA}^{-1}\mathbf{B}(1-\gamma)\gamma T(\theta_2 - \theta_1)$$
$$+\mathbf{CA}^{-1}\left[\left(\mathbf{A}^{-1}(e^{\mathbf{A}\theta_1} - \mathbf{I}) - \theta_1\right)\rho + \left(\mathbf{A}^{-1}(e^{\mathbf{A}\theta_2} - \mathbf{I}) - \theta_2\right)\eta\right.$$
$$\left.+\mathbf{A}^{-2}(e^{\mathbf{A}\theta_1} - \mathbf{I} - e^{\mathbf{A}\theta_2} + \mathbf{I})\mathbf{B} - \mathbf{A}^{-1}\mathbf{B}(\theta_1 - \theta_2)\right]\Big\} \tag{10.45}$$

We replace ρ and η in (10.45) with formulas (10.5) and (10.4),

$$y_0 = \tfrac{1}{T}\left\{y(0)(\theta_1 - \theta_2) - \mathbf{CA}^{-1}\mathbf{B}(1-\gamma)\gamma T(\theta_1 - \theta_2)\right.$$
$$+\mathbf{CA}^{-1}\left[\left(\mathbf{A}^{-1}(e^{\mathbf{A}\theta_1} - \mathbf{I}) - \theta_1\right)\mathbf{A}^{-1}(\mathbf{I} - e^{\mathbf{A}T})^{-1}[e^{\mathbf{A}T} - 2e^{\mathbf{A}\theta_2} + \mathbf{I}]\mathbf{B}\right.$$
$$+ \left(\mathbf{A}^{-1}(e^{\mathbf{A}\theta_2} - \mathbf{I}) - \theta_2\right)\mathbf{A}^{-1}(\mathbf{I} - e^{\mathbf{A}T})^{-1}[2e^{\mathbf{A}\theta_1} - e^{\mathbf{A}T} - \mathbf{I}]\mathbf{B}$$
$$\left.\left. +\mathbf{A}^{-2}(e^{\mathbf{A}\theta_1} - e^{\mathbf{A}\theta_2})\mathbf{B} - \mathbf{A}^{-1}\mathbf{B}(\theta_1 - \theta_2)\right]\right\}$$
$$= \tfrac{1}{T}\left\{y(0)(\theta_1 - \theta_2) - \mathbf{CA}^{-1}\mathbf{B}(1-\gamma)\gamma T(\theta_1 - \theta_2)\right.$$
$$+\mathbf{CA}^{-1}\left[\mathbf{A}^{-2}(e^{\mathbf{A}\theta_1} - \mathbf{I})(\mathbf{I} - e^{\mathbf{A}T})^{-1}[e^{\mathbf{A}T} - 2e^{\mathbf{A}\theta_2} + \mathbf{I}]\right.$$
$$-\mathbf{A}^{-1}(\mathbf{I} - e^{\mathbf{A}T})^{-1}[e^{\mathbf{A}T} - 2e^{\mathbf{A}\theta_2} + \mathbf{I}]\theta_1$$
$$+\mathbf{A}^{-2}(e^{\mathbf{A}\theta_2} - \mathbf{I})(\mathbf{I} - e^{\mathbf{A}T})^{-1}[2e^{\mathbf{A}\theta_1} - e^{\mathbf{A}T} - \mathbf{I}]$$
$$-\mathbf{A}^{-1}(\mathbf{I} - e^{\mathbf{A}T})^{-1}[2e^{\mathbf{A}\theta_1} - e^{\mathbf{A}T} - \mathbf{I}]\theta_2$$
$$\left.\left. +\mathbf{A}^{-2}(e^{\mathbf{A}\theta_1} - e^{\mathbf{A}\theta_2}) - \mathbf{A}^{-1}(\theta_1 - \theta_2)\right]\mathbf{B}\right\}.$$
(10.46)

We substitute for θ_1 and θ_2 (except for powers of exponents),

$$y_0 = \tfrac{1}{T}\left\{y(0)(2\gamma - 1)T - \mathbf{CA}^{-1}\mathbf{B}(1-\gamma)\gamma T(2\gamma - 1)T\right.$$
$$+\mathbf{CA}^{-2}\left[\mathbf{A}^{-1}(e^{\mathbf{A}\theta_1} - \mathbf{I})(\mathbf{I} - e^{\mathbf{A}T})^{-1}(e^{\mathbf{A}T} + \mathbf{I})\right.$$
$$-\mathbf{A}^{-1}(e^{\mathbf{A}\theta_1} - \mathbf{I})(\mathbf{I} - e^{\mathbf{A}T})^{-1}2e^{\mathbf{A}\theta_2}$$
$$-(\mathbf{I} - e^{\mathbf{A}T})^{-1}(e^{\mathbf{A}T} + \mathbf{I})\gamma T + (\mathbf{I} - e^{\mathbf{A}T})^{-1}2e^{\mathbf{A}\theta_2}\gamma T$$
$$-\mathbf{A}^{-1}(e^{\mathbf{A}\theta_2} - \mathbf{I})(\mathbf{I} - e^{\mathbf{A}T})^{-1}(e^{\mathbf{A}T} + \mathbf{I})$$
$$+\mathbf{A}^{-1}(e^{\mathbf{A}\theta_2} - \mathbf{I})(\mathbf{I} - e^{\mathbf{A}T})^{-1}2e^{\mathbf{A}\theta_1}$$
$$+(\mathbf{I} - e^{\mathbf{A}T})^{-1}(e^{\mathbf{A}T} + \mathbf{I})(1-\gamma)T$$
$$-(\mathbf{I} - e^{\mathbf{A}T})^{-1}2e^{\mathbf{A}\theta_1}(1-\gamma)T$$
$$\left.\left. +\mathbf{A}^{-1}(e^{\mathbf{A}\theta_1} - e^{\mathbf{A}\theta_2}) - (2\gamma - 1)T\right]\mathbf{B}\right\}.$$
(10.47)

Rearranging the terms in (10.47), we obtain

$$y_0 = \tfrac{1}{T}\left\{y(0)(2\gamma - 1)T - \mathbf{CA}^{-1}\mathbf{B}(1-\gamma)\gamma T(2\gamma - 1)T\right.$$
$$+\mathbf{CA}^{-2}\left[\mathbf{A}^{-1}(e^{\mathbf{A}\theta_1} - e^{\mathbf{A}\theta_2})(\mathbf{I} - e^{\mathbf{A}T})^{-1}(e^{\mathbf{A}T} + \mathbf{I})\right.$$
$$-2\mathbf{A}^{-1}(e^{\mathbf{A}T} - e^{\mathbf{A}\theta_2} - e^{\mathbf{A}T} + e^{\mathbf{A}\theta_1})(\mathbf{I} - e^{\mathbf{A}T})^{-1}$$
$$-(\mathbf{I} - e^{\mathbf{A}T})^{-1}(e^{\mathbf{A}T} + \mathbf{I})(2\gamma - 1)T$$
$$+2(\mathbf{I} - e^{\mathbf{A}T})^{-1}(\gamma e^{\mathbf{A}\theta_2} - (1-\gamma)e^{\mathbf{A}\theta_1})T$$
$$\left.\left. +\mathbf{A}^{-1}(e^{\mathbf{A}\theta_1} - e^{\mathbf{A}\theta_2}) - (2\gamma - 1)T\right]\mathbf{B}\right\}.$$
(10.48)

10.3 The LPRS derivation for an integrating linear part

Simplifying (10.48), we obtain

$$y_0 = \tfrac{1}{T}\left\{y(0)(2\gamma-1)T - \mathbf{CA}^{-1}\mathbf{B}(1-\gamma)\gamma T^2(2\gamma-1)\right.$$
$$+\mathbf{CA}^{-2}\left[\mathbf{A}^{-1}(e^{\mathbf{A}\theta_1}-e^{\mathbf{A}\theta_2})(\mathbf{I}-e^{\mathbf{A}T})^{-1}(e^{\mathbf{A}T}+\mathbf{I})\right.$$
$$-2\mathbf{A}^{-1}(e^{\mathbf{A}\theta_1}-e^{\mathbf{A}_2})(\mathbf{I}-e^{\mathbf{A}T})^{-1}$$
$$-(\mathbf{I}-e^{\mathbf{A}T})^{-1}(e^{\mathbf{A}T}+\mathbf{I})(2\gamma-1)T \qquad (10.49)$$
$$+2(\mathbf{I}-e^{\mathbf{A}T})^{-1}(\gamma(e^{\mathbf{A}\theta_2}+e^{\mathbf{A}\theta_1})-e^{\mathbf{A}\theta_1})T$$
$$\left.\left.+\mathbf{A}^{-1}(e^{\mathbf{A}\theta_1}-e^{\mathbf{A}\theta_2})-(2\gamma-1)T\right]\mathbf{B}\right\}.$$

Now we compute the following limit, which is instrumental below,

$$\lim_{\gamma\to\frac{1}{2}}\frac{\gamma(e^{\mathbf{A}\theta_1}-e^{\mathbf{A}\theta_2})-e^{\mathbf{A}\theta_1}}{2\gamma-1} = \lim_{\gamma\to\frac{1}{2}}\frac{\gamma(e^{\mathbf{A}\gamma T}-e^{\mathbf{A}(1-\gamma)T})-e^{\mathbf{A}\gamma T}}{2\gamma-1}$$
$$= \lim_{\gamma\to\frac{1}{2}}\frac{(\gamma-1)e^{\mathbf{A}\gamma T}+\gamma e^{\mathbf{A}(1-\gamma)T}}{2\gamma-1}.$$

Taking derivatives of the numerator and the denominator, we find the limit

$$\lim_{\gamma\to\frac{1}{2}}\frac{\gamma(e^{\mathbf{A}\theta_1}-e^{\mathbf{A}\theta_2})-e^{\mathbf{A}\theta_1}}{2\gamma-1}$$
$$= \lim_{\gamma\to\frac{1}{2}}\frac{e^{\mathbf{A}\gamma T}+(\gamma-1)\mathbf{A}Te^{\mathbf{A}\gamma T}+e^{\mathbf{A}(1-\gamma)T}-\gamma\mathbf{A}Te^{\mathbf{A}(1-\gamma)T}}{2} \qquad (10.50)$$
$$= \tfrac{1}{2}\left(e^{\mathbf{A}T/2}-\tfrac{1}{2}\mathbf{A}Te^{\mathbf{A}T/2}+e^{\mathbf{A}T/2}-\tfrac{1}{2}\mathbf{A}Te^{\mathbf{A}T/2}\right)$$
$$= \tfrac{1}{2}\left(2e^{\mathbf{A}T/2}-\mathbf{A}Te^{\mathbf{A}T/2}\right) = \left(\mathbf{I}-\tfrac{1}{2}\mathbf{A}T\right)e^{\mathbf{A}T/2}.$$

Considering formula (10.49) and auxiliary limits (10.11) and (10.50), we find that limit $\lim_{u_0\to 0}\frac{y_0}{u_0}$

$$\lim_{u_0\to 0}\frac{y_0}{u_0} = \lim_{\gamma\to\frac{1}{2}}\frac{y_0}{2\gamma-1} = \tfrac{1}{T}\left\{y(0)T-\tfrac{1}{4}\mathbf{CA}^{-1}\mathbf{B}T^2\right.$$
$$+\mathbf{CA}^{-2}\left[Te^{\mathbf{A}T/2}(\mathbf{I}-e^{\mathbf{A}T})^{-1}(e^{\mathbf{A}T}+\mathbf{I})\right.$$
$$-2Te^{\mathbf{A}T/2}(\mathbf{I}-e^{\mathbf{A}T})^{-1}-(\mathbf{I}-e^{\mathbf{A}T})^{-1}(e^{\mathbf{A}T}+\mathbf{I})T$$
$$\left.\left.+2(\mathbf{I}-e^{\mathbf{A}T})^{-1}\left(\mathbf{I}-\tfrac{1}{2}\mathbf{A}T\right)e^{\mathbf{A}T/2}T+Te^{\mathbf{A}T/2}-T\right]\mathbf{B}\right\}. \qquad (10.51)$$

We substitute formula (10.42) for $y(0)$:

$$\lim_{u_0\to 0}\frac{y_0}{u_0} = \tfrac{1}{4}T\mathbf{CA}^{-1}\mathbf{B} - \tfrac{1}{4}T\mathbf{CA}^{-1}\mathbf{B}$$
$$+\mathbf{CA}^{-2}\left[\tfrac{1}{2}(\mathbf{I}-e^{\mathbf{A}T})^{-1}\left(3e^{\mathbf{A}T}-3e^{\mathbf{A}T/2}-e^{\mathbf{A}T}e^{\mathbf{A}T/2}+\mathbf{I}\right)\right.$$
$$-\tfrac{1}{2}e^{\mathbf{A}T/2}+\tfrac{1}{2}\mathbf{I}+e^{\mathbf{A}T/2}(\mathbf{I}-e^{\mathbf{A}T})^{-1}(e^{\mathbf{A}T}+\mathbf{I})-2e^{\mathbf{A}T/2}(\mathbf{I}-e^{\mathbf{A}T})^{-1}$$
$$-(\mathbf{I}-e^{\mathbf{A}T})^{-1}(e^{\mathbf{A}T}+\mathbf{I})+2(\mathbf{I}-e^{\mathbf{A}T})^{-1}\left(\mathbf{I}-\tfrac{1}{2}\mathbf{A}T\right)e^{\mathbf{A}T/2}$$
$$\left.+e^{\mathbf{A}T/2}-\mathbf{I}\right]\mathbf{B}.$$
$$(10.52)$$

Further simplification yields the following formula:

$$\lim_{u_0 \to 0} \frac{y_0}{u_0} = \mathbf{CA}^{-2} \left[(\mathbf{I} - e^{\mathbf{A}T})^{-1} \left(\tfrac{3}{2} e^{\mathbf{A}T} - \tfrac{3}{2} e^{\mathbf{A}T/2} - \tfrac{1}{2} e^{\mathbf{A}T} e^{\mathbf{A}T/2} + \tfrac{1}{2} \mathbf{I} \right. \right.$$
$$\left. + e^{\mathbf{A}T} e^{\mathbf{A}T/2} + e^{\mathbf{A}T/2} - 2 e^{\mathbf{A}T/2} - e^{\mathbf{A}T} - \mathbf{I} + 2 e^{\mathbf{A}T/2} - \mathbf{A}T e^{\mathbf{A}T/2} \right)$$
$$\left. + \tfrac{1}{2} e^{\mathbf{A}T/2} - \tfrac{1}{2} \mathbf{I} \right] \mathbf{B}$$
$$= \mathbf{CA}^{-2} \left[(\mathbf{I} - e^{\mathbf{A}T})^{-1} \left(\tfrac{1}{2} e^{\mathbf{A}T} - \left(\tfrac{1}{2} \mathbf{I} + \mathbf{A}T \right) e^{\mathbf{A}T/2} + \tfrac{1}{2} e^{\mathbf{A}T} e^{\mathbf{A}T/2} - \tfrac{1}{2} \mathbf{I} \right) \right.$$
$$\left. + \tfrac{1}{2} e^{\mathbf{A}T/2} - \tfrac{1}{2} \mathbf{I} \right] \mathbf{B}.$$
(10.53)

Considering the LPRS definition, we write the formula for the real part:

$$\mathrm{Re} J(\omega) = -0.5 \lim_{f_0 \to 0} \frac{\sigma_0}{u_0} = 0.5 \lim_{\gamma \to \frac{1}{2}} \frac{y_0}{u_0}$$
$$= \tfrac{1}{4} \mathbf{CA}^{-2} \left[(\mathbf{I} - e^{\mathbf{A}T})^{-1} \left(e^{\mathbf{A}T} - (\mathbf{I} + 2\mathbf{A}T) e^{\mathbf{A}T/2} + e^{\mathbf{A}T} e^{\mathbf{A}T/2} - \mathbf{I} \right) \right.$$
$$\left. + e^{\mathbf{A}T/2} - \mathbf{I} \right] \mathbf{B}.$$
(10.54)

The real part of the LPRS is derived under the assumption that the limits at $u_0 \to 0$ and at $\gamma \to \tfrac{1}{2}$ are equal. This is true only if the period T does not change. Therefore, we need to prove that this is the case. The frequency of the oscillations (or the period) is defined by the switching condition,

$$y(0) = -y(\theta_1) = -b.$$

We find the derivative $\frac{\partial y(0)}{\partial u_0}$ at the point $\gamma = \tfrac{1}{2}$ by taking the derivative of (10.41),

$$\frac{\partial y(0)}{\partial u_0} = \tfrac{1}{2} \frac{\partial y(0)}{\partial \gamma} = (1 - 2\gamma) T \mathbf{C} \mathbf{A}^{-1} \mathbf{B} + \tfrac{1}{4} \mathbf{C} \mathbf{A}^{-2}$$
$$\times \left\{ (\mathbf{I} - e^{\mathbf{A}T})^{-1} \left[-3\mathbf{A}T(e^{\mathbf{A}\gamma T} - e^{\mathbf{A}(1-\gamma)T}) - e^{\mathbf{A}T} \mathbf{A}T(e^{\mathbf{A}\gamma T} - e^{\mathbf{A}(1-\gamma)T}) \right] \right.$$
$$\left. - \mathbf{A}T(e^{\mathbf{A}\gamma T} - e^{\mathbf{A}(1-\gamma)T}) \right\} \mathbf{B}.$$

or continuing

$$\left. \frac{\partial y(0)}{\partial u_0} \right|_{u_0 = 0} = \tfrac{1}{2} \left. \frac{\partial y(0)}{\partial \gamma} \right|_{\gamma = 1/2} = 0 \cdot T \mathbf{C} \mathbf{A}^{-1} \mathbf{B} + \tfrac{1}{4} \mathbf{C} \mathbf{A}^{-2}$$
$$\times \left\{ (\mathbf{I} - e^{\mathbf{A}T})^{-1} \left[-3\mathbf{A}T \cdot 0 - e^{\mathbf{A}T} \mathbf{A}T \cdot 0 \right] - \mathbf{A}T \cdot 0 \right\} \mathbf{B} = 0.$$
(10.55)

The derivative (10.55) is zero, which is also equivalent to $\left. \frac{\partial \Omega}{\partial f_0} \right|_{f_0 = 0} = 0$; hence the two limits are equal. Finally, the LPRS for the case of an integrating linear part can be written through the following formula,

$$J(\omega) = 0.25 \mathbf{C} \mathbf{A}^{-2} \{ (\mathbf{I} - \mathbf{D}^2)^{-1} [\mathbf{D}^2 - (\mathbf{I} + \tfrac{4\pi}{\omega} \mathbf{A}) \mathbf{D} + \mathbf{D}^3 - \mathbf{I}] + \mathbf{D} - \mathbf{I} \} \mathbf{B}$$
$$+ j \tfrac{\pi}{8} \mathbf{C} \mathbf{A}^{-1} \{ \tfrac{\pi}{\omega} + \mathbf{A}^{-1} [(\mathbf{I} - \mathbf{D}^2)^{-1} \cdot (3\mathbf{D}^2 - 3\mathbf{D} - \mathbf{D}^3 + \mathbf{I}) - \mathbf{D} + \mathbf{I}] \} \mathbf{B},$$
(10.56)

where $\mathbf{D} = e^{\frac{\pi}{\omega}A}$.

Therefore, the state-space description–based LPRS formula for the case of an integrating linear part (*type 1 relay servo system*) has been derived above.

10.4 Orbital stability of a system with an integrating linear part

We shall assess the orbital stability of the relay feedback system with an integrating plant by finding the Poincaré return map for the deviation of the variable $\mathbf{x}(t)$ at the switching times $\delta\rho \to \delta\eta \to \delta\rho$. The exclusion of the variable $y(t)$ from this consideration is possible, since the values of $y(t)$ at the switching times are confined by the switching condition $y = \pm b$. A similar exclusion of one of the state variables from consideration is also possible in the analysis of stability for the system with a non-integrating linear part. However, technically the consideration of the full-dimensional state vector leads to simpler derivations.

Therefore, let the periodic solution be given by (10.4) and (10.5), where $\rho_p = \mathbf{x}(0)$ and $\eta_p = \mathbf{x}(\theta_1)$, and the switching conditions $y(0) = -y(\theta_1) = -b$, with the constant external input being $f_0 \equiv (2\gamma - 1)\mathbf{C}\mathbf{A}^{-1}\mathbf{B}$. Let us assume that at time $t = 0$, the state vector is $\mathbf{x}(0) = \rho = \rho_p + \delta\rho$. Then for the time interval $t \in [0; t^*]$, where t^* is the time of the switch of the relay from "+1" to "−1," the state vector (while the control is $u = 1$) is given as follows:

$$\begin{aligned}\mathbf{x}(t) &= e^{\mathbf{A}t}(\rho_p + \delta\rho) + \mathbf{A}^{-1}(e^{\mathbf{A}t} - \mathbf{I})\mathbf{B} \\ &= e^{\mathbf{A}t}\rho_p + \mathbf{A}^{-1}(e^{\mathbf{A}t} - \mathbf{I})\mathbf{B} + e^{\mathbf{A}t}\delta\rho,\end{aligned} \quad (10.57)$$

where the first two addends give the unperturbed motion, and the third addend gives the motion due to the initial perturbation $\delta\rho$. Let us denote

$$\mathbf{x}(t^*) = \eta = \eta_p + \delta\eta. \quad (10.58)$$

We note that η_p is determined not at time $t = t^*$ but at time $t = \theta_1$.

The main task of our analysis is to find the mapping $\delta\rho \to \delta\eta$. Therefore, it follows from (10.58) that

$$\delta\eta = \mathbf{x}(t^*) - \eta_p. \quad (10.59)$$

Considering that all perturbations are small and the times t^* and θ_1 are close, we evaluate $\mathbf{x}(t^*)$ through a linear approximation of $\mathbf{x}(t)$ at the point $t = \theta_1$,

$$\mathbf{x}(t^*) - \mathbf{x}(\theta_1) = \dot{\mathbf{x}}(\theta_1-)(t^* - \theta_1),$$

where $\dot{\mathbf{x}}(\theta_1-)$ is the value of the derivative at the time immediately preceding time $t = \theta_1$. We express $\mathbf{x}(t^*)$ from the last equation as follows:

$$\mathbf{x}(t^*) = \mathbf{x}(\theta_1) + \dot{\mathbf{x}}(\theta_1-)(t^* - \theta_1). \quad (10.60)$$

Now we evaluate $\mathbf{x}(\theta_1)$ using (10.57):

$$\mathbf{x}(\theta_1) = \eta_p + e^{\mathbf{A}\theta_1}\delta\rho. \qquad (10.61)$$

We substitute (10.60) and (10.61) in (10.59):

$$\begin{aligned}\delta\eta &= \mathbf{x}(\theta_1) + \dot{\mathbf{x}}(\theta_1-)(t^* - \theta_1) - \eta_p \\ &= \eta_p + e^{\mathbf{A}\theta_1}\delta\rho + \dot{\mathbf{x}}(\theta_1-)(t^* - \theta_1) - \eta_p \\ &= \mathbf{v}(\theta_1-)(t^* - \theta_1) + e^{\mathbf{A}\theta_1}\delta\rho,\end{aligned} \qquad (10.62)$$

where $\mathbf{v}(t) = \dot{\mathbf{x}}(t)$. We evaluate $(t^* - \theta_1)$. The system output $y(t)$ crosses the level b at $t = t^*$, which results in the switch. Using linear approximation, we can write the following expression for the perturbation of the output:

$$(t^* - \theta_1)\dot{y}(\theta_1-) = -\delta y(\theta_1). \qquad (10.63)$$

Note that due to the integrating character of the plant, the following equality holds: $\dot{y}(\theta_1-) = \dot{y}(\theta_1+) = \dot{y}(\theta_1)$. Consider the equation

$$\dot{y} = \mathbf{C}\mathbf{x} - f_0$$

from which we obtain the following formula:

$$y(t) = \int_0^t \mathbf{C}\mathbf{x}(\tau)d\tau - \int_0^t f_0 d\tau + y(0),$$

where $y(0) = -b$ and $\int_0^t f_0 d\tau = f_0 t\big|_0^t = f_0 t$. Integration of the first addend yields

$$\begin{aligned}\int_0^t \mathbf{C}\mathbf{x}(\tau)d\tau &= \int_0^t \mathbf{C}\{e^{\mathbf{A}\tau}\rho_p + \mathbf{A}^{-1}(e^{\mathbf{A}\tau} - \mathbf{I})\mathbf{B} + e^{\mathbf{A}\tau}\delta\rho\}d\tau \\ &= \mathbf{C}\mathbf{A}^{-1}e^{\mathbf{A}t}\rho_p\big|_0^t + \mathbf{C}\mathbf{A}^{-2}e^{\mathbf{A}t}\mathbf{B}\big|_0^t - \mathbf{C}\mathbf{A}^{-1}\mathbf{B}t + \mathbf{C}\mathbf{A}^{-1}e^{\mathbf{A}t}\delta\rho\big|_0^t \\ &= \mathbf{C}\mathbf{A}^{-1}(e^{\mathbf{A}t} - \mathbf{I})\rho_p + \mathbf{C}\mathbf{A}^{-2}(e^{\mathbf{A}t} - \mathbf{I})\mathbf{B} - \mathbf{C}\mathbf{A}^{-1}\mathbf{B}t + \mathbf{C}\mathbf{A}^{-1}(e^{\mathbf{A}t} - \mathbf{I})\delta\rho.\end{aligned} \qquad (10.64)$$

We write a formula for $y(t)$ using the expression (10.64),

$$\begin{aligned}y(t) &= \mathbf{C}\mathbf{A}^{-1}(e^{\mathbf{A}t} - \mathbf{I})\rho_p + \mathbf{C}\mathbf{A}^{-2}(e^{\mathbf{A}t} - \mathbf{I})\mathbf{B} \\ &\quad - \mathbf{C}\mathbf{A}^{-1}\mathbf{B}t + \mathbf{C}\mathbf{A}^{-1}(e^{\mathbf{A}t} - \mathbf{I})\delta\rho - f_0 t - b.\end{aligned} \qquad (10.65)$$

The component that is due to the initial perturbation is the one that contains $\delta\rho$. Therefore, the perturbation of the output at time $t = \theta_1$ is

$$\delta y(\theta_1) = \mathbf{C}\mathbf{A}^{-1}(e^{\mathbf{A}\theta_1} - \mathbf{I})\delta\rho. \qquad (10.66)$$

10.4 Orbital stability of a system with an integrating linear part

We evaluate $\dot{y}(\theta_1)$ in formula (10.63) as

$$\dot{y} = \mathbf{C}\mathbf{x} - f_0$$

$$\dot{y}(\theta_1) = \mathbf{C}\mathbf{x}(\theta_1) - f_0 = \mathbf{C}\eta_p + \mathbf{C}e^{\mathbf{A}\theta_1}\delta\rho - f_0. \tag{10.67}$$

We can now write an expression for $(t^* - \theta_1)$ as follows:

$$t^* - \theta_1 = -\frac{\delta y(\theta_1)}{\dot{y}(\theta_1)} = -\frac{\mathbf{C}\mathbf{A}^{-1}(e^{\mathbf{A}\theta_1} - \mathbf{I})\delta\rho}{\dot{y}(\theta_1)}. \tag{10.68}$$

We substitute (10.68) for $(t^* - \theta_1)$ in (10.62):

$$\delta\eta = -\mathbf{v}(\theta_1-)\frac{\delta y(\theta_1)}{\dot{y}(\theta_1)} = -\mathbf{v}(\theta_1-)\frac{\mathbf{C}\mathbf{A}^{-1}(e^{\mathbf{A}\theta_1}-\mathbf{I})\delta\rho}{\dot{y}(\theta_1)} + e^{\mathbf{A}\theta_1}\delta\rho$$

$$= -\mathbf{v}(\theta_1-)\frac{\mathbf{C}\mathbf{A}^{-1}(e^{\mathbf{A}\theta_1}-\mathbf{I})}{\mathbf{C}\eta_p + \mathbf{C}e^{\mathbf{A}\theta_1}\delta\rho - f_0}\delta\rho + e^{\mathbf{A}\theta_1}\delta\rho.$$

The Jacobian matrix for the mapping $\delta\rho \to \delta\eta$ can be found by taking the derivative of the previous formula with respect to $\delta\rho$ at the point $\delta\rho = \mathbf{0}$,

$$\Phi_1 = \left.\frac{d(\delta\eta)}{d(\delta\rho)}\right|_{\delta\rho=\mathbf{0}} = -\mathbf{v}(\theta_1-)\frac{\mathbf{C}\mathbf{A}^{-1}(e^{\mathbf{A}\theta_1}-\mathbf{I})}{\dot{y}_p(\theta_1)} + e^{\mathbf{A}\theta_1}, \tag{10.69}$$

where $\mathbf{v}(\theta_1-)$ is given by formula (10.26), $\dot{y}_p(\theta_1) = \mathbf{C}\eta_p - f_0$, and η_p is given by (10.4). Therefore,

$$\dot{y}_p(\theta_1) = \mathbf{C}(\mathbf{I} - e^{\mathbf{A}T})^{-1}\mathbf{A}^{-1}[2e^{\mathbf{A}\theta_1} - e^{\mathbf{A}T} - \mathbf{I}]\mathbf{B} - f_0. \tag{10.70}$$

To be able to assess the local orbital stability of the system, we need to find the Jacobian matrix of the Poincare return map, which arises from the chain-rule application of two mappings: $\delta\rho \to \delta\eta$ and $\delta\eta \to \delta\rho$. Reasoning along the same lines, we find the Jacobian matrix of mapping $\delta\eta \to \delta\rho$,

$$\Phi_2 = -\mathbf{v}(\theta_2-)\frac{\mathbf{C}\mathbf{A}^{-1}(e^{\mathbf{A}\theta_2}-\mathbf{I})}{\dot{y}_p(\theta_2)} + e^{\mathbf{A}\theta_2}, \tag{10.71}$$

where $\mathbf{v}(\theta_2-)$ is given by formula (10.27), $\dot{y}_p(\theta_2) = \mathbf{C}\rho_p - f_0$, and ρ_p is given by (10.5). Therefore,

$$\dot{y}_p(\theta_2) = -\mathbf{C}(\mathbf{I} - e^{\mathbf{A}T})^{-1}\mathbf{A}^{-1}[2e^{\mathbf{A}\theta_2} - e^{\mathbf{A}T} - \mathbf{I}]\mathbf{B} - f_0. \tag{10.72}$$

The local orbital stability of the system is determined by the eigenvalues of the matrix $\Phi = \Phi_2\Phi_1$. If all eigenvalues of Φ have magnitudes less than *one*, then the system is orbitally asymptotically stable.

For the symmetric motion, it is sufficient to check only half of the period of motion. If, therefore, all eigenvalues of the matrix

$$\Phi_0 = -\frac{\mathbf{v}\left(\frac{T}{2}-\right)\mathbf{C}\mathbf{A}^{-1}(e^{\mathbf{A}\frac{T}{2}} - \mathbf{I})}{\dot{y}_p(\frac{T}{2})} + e^{\mathbf{A}\frac{T}{2}}, \tag{10.73}$$

where

$$\mathbf{v}(\frac{T}{2}-) = 2\left(\mathbf{I} - e^{\mathbf{A}T}\right)^{-1}\left(e^{\mathbf{A}\frac{T}{2}} - e^{\mathbf{A}T}\right)\mathbf{B} = 2\left(\mathbf{I} + e^{\mathbf{A}T/2}\right)^{-1} e^{\mathbf{A}T/2}\mathbf{B},$$

and

$$\begin{aligned}
\dot{y}_p\left(\tfrac{T}{2}\right) &= \mathbf{C}\left(\mathbf{I} - e^{\mathbf{A}T}\right)^{-1}\mathbf{A}^{-1}\left(2e^{\mathbf{A}T/2} - e^{\mathbf{A}T} - \mathbf{I}\right)\mathbf{B} \\
&= \mathbf{C}\mathbf{A}^{-1}\left(\mathbf{I} - e^{\mathbf{A}T}\right)^{-1}\left(2e^{\mathbf{A}T/2} - e^{\mathbf{A}T} - \mathbf{I}\right)\mathbf{B} \\
&= \mathbf{C}\mathbf{A}^{-1}\left(\mathbf{I} - e^{\mathbf{A}T}\right)^{-1}\left(2e^{\mathbf{A}T/2} + \mathbf{I} - e^{\mathbf{A}T} - 2\mathbf{I}\right)\mathbf{B} \\
&= \mathbf{C}\mathbf{A}^{-1}\left(\mathbf{I} - e^{\mathbf{A}T}\right)^{-1}\left(2\left(e^{\mathbf{A}T/2} - \mathbf{I}\right) + \left(\mathbf{I} - e^{\mathbf{A}T}\right)\right)\mathbf{B} \\
&= \mathbf{C}\mathbf{A}^{-1}\mathbf{B} + 2\mathbf{C}\mathbf{A}^{-1}\left(\mathbf{I} - e^{\mathbf{A}T}\right)^{-1}\left(e^{\mathbf{A}T/2} - \mathbf{I}\right)\mathbf{B} \\
&= \mathbf{C}\mathbf{A}^{-1}\mathbf{B} - 2\mathbf{C}\mathbf{A}^{-1}\left(\mathbf{I} - e^{\mathbf{A}T/2}\right)^{-1}\left(\mathbf{I} + e^{\mathbf{A}T/2}\right)^{-1}\left(\mathbf{I} - e^{\mathbf{A}T/2}\right)\mathbf{B} \\
&= \mathbf{C}\mathbf{A}^{-1}\mathbf{B} - 2\mathbf{C}\mathbf{A}^{-1}\left(\mathbf{I} + e^{\mathbf{A}T/2}\right)^{-1}\mathbf{B}
\end{aligned} \quad (10.74)$$

have magnitudes less than *one*, then the system is locally orbitally asymptotically stable.

10.5 The LPRS derivation for a linear part with time delay

Let the plant be

$$\begin{aligned} \dot{x} &= \mathbf{A}x + \mathbf{B}u \\ y &= \mathbf{C}x \end{aligned} \quad (10.75)$$

and the control be

$$u = \begin{cases} +1 \text{ if } \sigma(t-\tau) = f_0 - y(t-\tau) \geq b & \text{or} \quad \sigma(t-\tau) > -b, \ u(t-) = 1 \\ -1 \text{ if } \sigma(t-\tau) = f_0 - y(t-\tau) \leq -b & \text{or} \quad \sigma(t-\tau) < b, \ u(t-) = -1 \end{cases}$$

where $\mathbf{A} \in R^{n \times n}$, $\mathbf{B} \in R^{n \times 1}$, $\mathbf{C} \in R^{1 \times n}$ are matrices, and $\mathbf{A}$ is nonsingular. We assume that time $t = 0$ corresponds to the time the error signal reaches the value of the hysteresis: $\sigma = b, \dot{\sigma} > 0$. The solution for the constant control $u = \pm 1$ is

$$\mathbf{x}(t) = e^{\mathbf{A}(t-\tau)}\mathbf{x}(\tau) \pm \mathbf{A}^{-1}(e^{\mathbf{A}(t-\tau)} - \mathbf{I})\mathbf{B}, \quad t > \tau.$$

Therefore, also

$$\mathbf{x}(\tau) = e^{\mathbf{A}\tau}\mathbf{x}(0) - \mathbf{A}^{-1}(e^{\mathbf{A}\tau} - \mathbf{I})\mathbf{B}.$$

A fixed point of the Poincaré return map for asymmetric periodic motion with positive pulse length θ_1, negative pulse length θ_2, and unity amplitude (the LPRS does not depend on the control amplitude) is determined as follows:

$$\eta_p = e^{\mathbf{A}\theta_1}\rho_p + \mathbf{A}^{-1}(e^{\mathbf{A}\theta_1} - \mathbf{I})\mathbf{B}, \quad (10.76)$$

10.5 The LPRS derivation for a linear part with time delay

$$\rho_p = e^{\mathbf{A}\theta_2}\eta_p - \mathbf{A}^{-1}(e^{\mathbf{A}\theta_2} - \mathbf{I})\mathbf{B}, \tag{10.77}$$

where $\rho_p = \mathbf{x}(\tau) = \mathbf{x}(T+\tau)$, $\eta_p = \mathbf{x}(\theta_1 + \tau)$. Suppose θ_1 and θ_2 are known. Then (10.76) and (10.77) for ρ_p and η_p are identical to (10.2) and (10.3) (however, the variables themselves are defined in a different way) and can be solved in exactly the same way:

$$\eta_p = (\mathbf{I} - e^{\mathbf{A}T})^{-1}\mathbf{A}^{-1}[2e^{\mathbf{A}\theta_1} - e^{\mathbf{A}T} - \mathbf{I}]\mathbf{B}, \tag{10.78}$$

$$\rho_p = (\mathbf{I} - e^{\mathbf{A}T})^{-1}\mathbf{A}^{-1}[e^{\mathbf{A}T} - 2e^{\mathbf{A}\theta_2} + \mathbf{I}]\mathbf{B}. \tag{10.79}$$

Now given that $\rho_p = \mathbf{x}(\tau) = e^{\mathbf{A}\tau}\mathbf{x}(0) - \mathbf{A}^{-1}(e^{\mathbf{A}\tau} - \mathbf{I})\mathbf{B}$ and $\eta_p = \mathbf{x}(\theta_1 + \tau) = e^{\mathbf{A}\tau}\mathbf{x}(\theta_1) + \mathbf{A}^{-1}(e^{\mathbf{A}\tau} - \mathbf{I})\mathbf{B}$, we find $\mathbf{x}(0)$:

$$\begin{aligned}\mathbf{x}(0) &= e^{-\mathbf{A}\tau}\left[\rho_p + \mathbf{A}^{-1}(e^{\mathbf{A}\tau} - \mathbf{I})\mathbf{B}\right] \\ &= e^{-\mathbf{A}\tau}\left[(\mathbf{I} - e^{\mathbf{A}T})^{-1}\mathbf{A}^{-1}[e^{\mathbf{A}T} - 2e^{\mathbf{A}\theta_2} + \mathbf{I}]\mathbf{B} + \mathbf{A}^{-1}(e^{\mathbf{A}\tau} - \mathbf{I})\mathbf{B}\right].\end{aligned} \tag{10.80}$$

Reasoning along similar lines, we find the formula for $\mathbf{x}(\theta_1)$,

$$\begin{aligned}\mathbf{x}(\theta_1) &= e^{-\mathbf{A}\tau}\left[\eta_p - \mathbf{A}^{-1}(e^{\mathbf{A}\tau} - \mathbf{I})\mathbf{B}\right] \\ &= e^{-\mathbf{A}\tau}\left[(\mathbf{I} - e^{\mathbf{A}T})^{-1}\mathbf{A}^{-1}[2e^{\mathbf{A}\theta_1} - e^{\mathbf{A}T} - \mathbf{I}]\mathbf{B} - \mathbf{A}^{-1}(e^{\mathbf{A}\tau} - \mathbf{I})\mathbf{B}\right].\end{aligned} \tag{10.81}$$

Consider now the symmetric motion as a limit of (10.80) at $\theta_1; \theta_2 \to \theta = T/2$,

$$\begin{aligned}\lim_{\theta_1;\theta_2 \to \theta = T/2} \mathbf{x}(0) &= e^{-\mathbf{A}\tau}\left[(\mathbf{I} - e^{\mathbf{A}T})^{-1}\mathbf{A}^{-1}[e^{\mathbf{A}T} - 2e^{\mathbf{A}T/2} + \mathbf{I}]\mathbf{B} \right.\\ &\quad \left. + \mathbf{A}^{-1}(e^{\mathbf{A}\tau} - \mathbf{I})\mathbf{B}\right] \\ &= e^{-\mathbf{A}\tau}\left[(\mathbf{I} - e^{\mathbf{A}T/2})^{-1}(\mathbf{I} + e^{\mathbf{A}T/2})^{-1}\mathbf{A}^{-1}(\mathbf{I} - e^{\mathbf{A}T/2})^2\mathbf{B} \right.\\ &\quad \left. + \mathbf{A}^{-1}(e^{\mathbf{A}\tau} - \mathbf{I})\mathbf{B}\right] \\ &= e^{-\mathbf{A}\tau}\left[(\mathbf{I} + e^{\mathbf{A}T/2})^{-1}\mathbf{A}^{-1}(\mathbf{I} - e^{\mathbf{A}T/2}) + \mathbf{A}^{-1}(e^{\mathbf{A}\tau} - \mathbf{I})\right]\mathbf{B} \\ &= \left[e^{-\mathbf{A}\tau}(\mathbf{I} + e^{\mathbf{A}T/2})^{-1}\mathbf{A}^{-1}(\mathbf{I} - e^{\mathbf{A}T/2}) + e^{-\mathbf{A}\tau}\mathbf{A}^{-1}(e^{\mathbf{A}\tau} - \mathbf{I})\right]\mathbf{B} \\ &= (\mathbf{I} + e^{\mathbf{A}T/2})^{-1}\mathbf{A}^{-1}\left[e^{-\mathbf{A}\tau} - e^{\mathbf{A}(T/2-\tau)} \right.\\ &\quad \left. + (\mathbf{I} + e^{\mathbf{A}T/2})(\mathbf{I} - e^{-\mathbf{A}\tau})\right]\mathbf{B} \\ &= (\mathbf{I} + e^{\mathbf{A}T/2})^{-1}\mathbf{A}^{-1}\left[\mathbf{I} + e^{\mathbf{A}T/2} - 2e^{\mathbf{A}(T/2-\tau)}\right]\mathbf{B}.\end{aligned} \tag{10.82}$$

The imaginary part of the LPRS can be obtained from (10.82) in accordance with its definition as follows:

$$\begin{aligned}\mathrm{Im}J(\omega) &= \tfrac{\pi}{4}\lim_{\theta_1;\theta_2 \to \theta = T/2}\mathbf{x}(0) \\ &= \tfrac{\pi}{4}\mathbf{C}(\mathbf{I} + e^{\mathbf{A}\pi/\omega})^{-1}(\mathbf{I} + e^{\mathbf{A}\pi/\omega} - 2e^{\mathbf{A}(\pi/\omega-\tau)})\mathbf{A}^{-1}\mathbf{B}.\end{aligned} \tag{10.83}$$

For deriving the expression of the real part of the LPRS, consider the periodic solution (10.78) and (10.79) as a result of the feedback action,

$$\begin{cases} f_0 - y(0) = b \\ f_0 - y(\theta_1) = -b. \end{cases} \tag{10.84}$$

Solving the set of equations (10.84) for f_0, we obtain

$$f_0 = \frac{y(0) + y(\theta_1)}{2}.$$

Hence, the constant term of the error signal $\sigma(t)$ is

$$\sigma_0 = f_0 - y_0 = \frac{y(0) + y(\theta_1)}{2} - y_0.$$

The real part of the LPRS formula is transformed into

$$\operatorname{Re} J(\omega) = -0.5 \lim_{\gamma \to \frac{1}{2}} \frac{0.5\left[y(0) + y(\theta_1)\right] - y_0}{u_0}, \tag{10.85}$$

where $\gamma = \frac{\theta_1}{\theta_1 + \theta_2} = \frac{\theta_1}{T}$.
Then $\theta_1 = \gamma T$, $\theta_2 = (1 - \gamma)T$, $u_0 = 2\gamma - 1$, and (10.85) can be rewritten as:

$$\operatorname{Re} J(\omega) = -0.5 \lim_{\gamma \to \frac{1}{2}} \frac{0.5 \mathbf{C}\left[\mathbf{x}(0) + \mathbf{x}(\theta_1)\right] - y_0}{2\gamma - 1}, \tag{10.86}$$

where $\mathbf{x}(0)$ and $\mathbf{x}(\theta_1)$ are given by (10.80) and (10.81), respectively. Now considering the limit given by (10.11), which can also be applied to the plant with a time delay, we find the following limit,

$$\begin{aligned}
\lim_{u_0 \to 0 (\theta_1 + \theta_2 = T = const)} \frac{\mathbf{x}(0) + \mathbf{x}(\theta_1)}{u_0} &= \lim_{\gamma \to \frac{1}{2}} \frac{\mathbf{x}(0) + \mathbf{x}(\theta_1)}{2\gamma - 1} \\
&= \lim_{\gamma \to \frac{1}{2}} \frac{e^{-\mathbf{A}\tau}(\rho_p + \eta_p)}{2\gamma - 1} \\
&= e^{-\mathbf{A}\tau} \lim_{\gamma \to \frac{1}{2}} \frac{2(\mathbf{I} - e^{\mathbf{A}T})^{-1} \mathbf{A}^{-1}[e^{\mathbf{A}\theta_1} - e^{\mathbf{A}\theta_2}]\mathbf{B}}{2\gamma - 1} \\
&= 2e^{-\mathbf{A}\tau}(\mathbf{I} - e^{\mathbf{A}T})^{-1} \mathbf{A}^{-1} \lim_{\gamma \to \frac{1}{2}} \frac{e^{\mathbf{A}\theta_1} - e^{\mathbf{A}\theta_2}}{2\gamma - 1} \mathbf{B} \\
&= 2e^{-\mathbf{A}\tau}(\mathbf{I} - e^{\mathbf{A}T})^{-1} \mathbf{A}^{-1} \mathbf{A} T e^{\mathbf{A}T/2} \mathbf{B} \\
&= 2e^{-\mathbf{A}\tau} T (\mathbf{I} - e^{\mathbf{A}T})^{-1} e^{\mathbf{A}T/2} \mathbf{B}.
\end{aligned} \tag{10.87}$$

To find the limit $\lim_{u_0 \to 0} \frac{y_0}{u_0}$, consider the equations for the constant terms of the variables (averaged variables), which are obtained from the original equations of the plant by equating the derivatives to zero,

$$\begin{cases} 0 = \mathbf{A}\mathbf{x}_0 + \mathbf{B}u_0 \\ y_0 = \mathbf{C}\mathbf{x}_0. \end{cases}$$

From these equations, we obtain $\mathbf{x}_0 = -\mathbf{A}^{-1}\mathbf{B}u_0$ and $y_0 = -\mathbf{C}\mathbf{A}^{-1}\mathbf{B}u_0$. Therefore,

$$\lim_{u_0 \to 0} \frac{y_0}{u_0} = -\mathbf{C}\mathbf{A}^{-1}\mathbf{B}, \tag{10.88}$$

10.5 The LPRS derivation for a linear part with time delay

which is essentially the steady-state gain of the plant. The real part of the LPRS is obtained by substituting (10.87) and (10.88) for the respective limits in (10.86):

$$\begin{aligned}
\mathrm{Re}\, J(\omega) &= -0.5 \lim_{u_0 \to 0} \frac{0.5\mathbf{C}[\mathbf{x}(0)+\mathbf{x}(\theta_1)]-y_0}{u_0} \\
&= -0.25\mathbf{C} \lim_{u_0 \to 0} \frac{\mathbf{x}(0)+\mathbf{x}(\theta_1)}{u_0} + 0.5 \lim_{u_0 \to 0} \frac{y_0}{u_0} \\
&= -0.5T\mathbf{C}(\mathbf{I}-e^{\mathbf{A}T})^{-1}e^{\mathbf{A}(T/2-\tau)}\mathbf{B} - 0.5\mathbf{C}\mathbf{A}^{-1}\mathbf{B}.
\end{aligned} \quad (10.89)$$

The real part of the LPRS is derived under the assumption that the limits at $u_0 \to 0$ and at $\gamma \to \frac{1}{2}$ are equal. This is true only if the period T does not change. Therefore, we need to prove that this is the case. The frequency of the oscillations (or the period) is defined by the following switching condition that is obtained from (10.82),

$$y(0) - y(\theta_1) = -2b.$$

We find the derivative $\frac{\partial(y(0)-y(\theta_1))}{\partial u_0} = \frac{1}{2}\frac{\partial(y(0)-y(\theta_1))}{\partial \gamma}$ at the point $\gamma = \frac{1}{2}$,

$$\begin{aligned}
y(0) - y(\theta_1) &= \mathbf{C}e^{-\mathbf{A}\tau}\{(\mathbf{I}-e^{\mathbf{A}T})^{-1}\mathbf{A}^{-1} \\
&\quad \times [e^{\mathbf{A}T}-2e^{\mathbf{A}\theta_2}+\mathbf{I}-2e^{\mathbf{A}\theta_1}+e^{\mathbf{A}T}+\mathbf{I}]\mathbf{B}+2\mathbf{A}^{-1}(e^{\mathbf{A}\tau}-\mathbf{I})\mathbf{B}\} \\
&= 2\mathbf{C}e^{-\mathbf{A}\tau}(\mathbf{I}-e^{\mathbf{A}T})^{-1}\mathbf{A}^{-1}\left[e^{\mathbf{A}T}+\mathbf{I}-(e^{\mathbf{A}\theta_1}+e^{\mathbf{A}\theta_2})\right]\mathbf{B} \\
&\quad + 2\mathbf{C}\mathbf{A}^{-1}(\mathbf{I}-e^{-\mathbf{A}\tau})\mathbf{B} \\
&= 2\mathbf{C}e^{-\mathbf{A}\tau}(\mathbf{I}-e^{\mathbf{A}T})^{-1}\mathbf{A}^{-1}\left[e^{\mathbf{A}T}+\mathbf{I}-(e^{\mathbf{A}\gamma T}+e^{\mathbf{A}(1-\gamma)T})\right]\mathbf{B} \\
&\quad + 2\mathbf{C}\mathbf{A}^{-1}(\mathbf{I}-e^{-\mathbf{A}\tau})\mathbf{B}.
\end{aligned}$$

We take the derivative with respect to γ in the last formula (considering also that $\mathbf{C}\mathbf{A}^{-1}(\mathbf{I}-e^{-\mathbf{A}\tau})\mathbf{B}$ does not depend on γ),

$$\begin{aligned}
\frac{\partial(y(0)-y(\theta_1))}{\partial \gamma} &= -2\mathbf{C}(\mathbf{I}-e^{\mathbf{A}T})^{-1}\mathbf{A}^{-1}\left[\mathbf{A}Te^{\mathbf{A}\gamma T}-\mathbf{A}Te^{\mathbf{A}(1-\gamma)T}\right]\mathbf{B} \\
&= -2\mathbf{C}(\mathbf{I}-e^{\mathbf{A}T})^{-1}T\left[e^{\mathbf{A}\gamma T}-e^{\mathbf{A}(1-\gamma)T}\right]\mathbf{B}.
\end{aligned}$$

The derivative $\frac{\partial(y(0)-y(\theta_1))}{\partial \gamma}$ at the point $\gamma = \frac{1}{2}$ is as follows,

$$\left.\frac{\partial(y(0)-y(\theta_1))}{\partial \gamma}\right|_{\gamma=1/2} = 0.$$

It follows from the last formula that the two limits are equivalent. It also follows from the last formula that

$$\left.\frac{d\Omega}{df_0}\right|_{f_0=0} = 0.$$

Finally, the state space description based formula of the LPRS can be obtained by combining formulas (10.81) and (10.89) as follows:

$$\begin{aligned}
J(\omega) &= -0.5\mathbf{C}\left[\mathbf{A}^{-1}+\frac{2\pi}{\omega}\left(\mathbf{I}-e^{\frac{2\pi}{\omega}\mathbf{A}}\right)^{-1}e^{\left(\frac{\pi}{\omega}-\tau\right)\mathbf{A}}\right]\mathbf{B} \\
&\quad + j\frac{\pi}{4}\mathbf{C}\left(\mathbf{I}+e^{\frac{\pi}{\omega}\mathbf{A}}\right)^{-1}\left(\mathbf{I}+e^{\frac{\pi}{\omega}\mathbf{A}}-2e^{\left(\frac{\pi}{\omega}-\tau\right)\mathbf{A}}\right)\mathbf{A}^{-1}\mathbf{B}.
\end{aligned} \quad (10.90)$$

10.6 MATLAB code for LPRS computing

```
function J=lprsmatr(A,B,C,w)
%
% Calculation of a point of the LPRS
% of a non-integrating plant
% for matrix-vector system description,
% dx/dt=Ax+Bu; y=Cx
% w - frequency
%
n=size(A,1);
AINV=inv(A);
I=eye(n);
if w==0
J=(-0.5+j*0.25*pi)*C*AINV*B;
else
t=2.*pi/w;
AEXP=expm(0.5*A*t);
AEXP2=expm(A*t);
re_lprs=-0.5*C*(AINV+t*inv(I-AEXP2)*AEXP)*B;
im_lprs=0.25*pi*C*inv(I+AEXP)*(I-AEXP)*AINV*B;
J=re_lprs+j*im_lprs;
end

function J=lprsmatrint(A,B,C,w)
%
% Calculation of a point of the LPRS
% of an integrating plant
% for matrix-vector system description,
% dx/dt=Ax+Bu; dy/dt=Cx
% w - frequency
%
n=size(A,1);
AINV=inv(A);
AINV2=AINV*AINV;
I=eye(n);
if w==0
J=0.5*C*AINV*B-j*1000000.;
else
t=2.*pi/w;
D=expm(0.5*A*t);
re_lprs=0.25*C*AINV2*(inv(I-D*D)*(D*D-(I+2.*t*A)*D+D*D*D-I)...
+D-I)*B;
im_lprs=0.0625*pi*C*AINV*B*t+0.125*pi*C*AINV*AINV*(inv(I-D*D)...
*(3*D*D-3*D-D*D*D+I)-D+I)*B;
```

10.6 MATLAB code for LPRS computing

```
J=re_lprs+j*im_lprs;
end

function J=lprsmatrdel(A,B,C,tau,w)
%
% Calculation of a point of the LPRS
% for matrix-vector system description having time delay "tau",
% dx/dt=Ax+Bu(t-tau); y=Cx
% w - frequency
%
n=size(A,1);
AINV=inv(A);
I=eye(n);
if w==0
J=(-0.5+j*0.25*pi)*C*AINV*B;
else
t=2.*pi/w;
AEXP=expm(0.5*A*t);
AEXP2=expm(A*t);
AEXP3=expm(A*(0.5*t-tau));
re_lprs=-0.5*C*(AINV+t*inv(I-AEXP2)*AEXP3)*B;
im_lprs=0.25*pi*C*inv(I+AEXP)*(I+AEXP-2*AEXP3)*AINV*B;
J=re_lprs+j*im_lprs;
end

function J=lprsser200(w,name,pr)
% Function calculating the LPRS
% at a given frequency
% based on the series formula
% (as a sum of 200 terms of the series)
% 'w' - current frequency,
% 'name' - name of m-file providing
% calculation of transfer function,
% 'name' is a string variable
% 'pr' - parameters of transfer function
reloc=0;
imloc=0;
iodd=-1;
for k=1:200
iodd=-iodd;
omk=k*w;
reimloc=feval(name,omk,pr);
reloc=reloc+iodd*real(reimloc);
if iodd==1
imloc=imloc+imag(reimloc)/k;
end
```

end
J=reloc+j*imloc;

function J=lprs1ord(k,t,w)
%
% Calculation of a point of the LPRS
% for Transfer Function G(s)=k/(t*s+1),
% w - frequency
%
if w==0
J=k*(0.5-j*pi/4);
else
al=pi/t/w;
J=0.5*k*(1-al*csch(al)-j*0.5*pi*tanh(al/2));
end

function J=lprsint(k,w)
%
% Calculation of a point of the LPRS
% for Transfer Function G(s)=k/s,
% w - frequency
%
if w==0
J=0-j*inf;
else
J=0-j*pi*pi*k/8/w;
end

function J=lprs2ord1(k,xi,w)
%
% Calculation of a point of the LPRS
% for Transfer Function G(s)=k/(s*s+2*xi*s+1),
% w - frequency
% xi < 1
%
if w==0
J=k*(0.5-j*pi/4);
else
al=pi*xi/w;
sq=sqrt(1-xi*xi);
bt=pi*sq/w;
gm=al/bt;
b=al*cos(bt)*sinh(al)+bt*sin(bt)*cosh(al);
c=al*sin(bt)*cosh(al)-bt*cos(bt)*sinh(al);
J=0.5*k*(1-(b+gm*c)/(sin(bt)∧2+sinh(al)∧2))...

```
-j*0.25*pi*k*(sinh(al)-gm*sin(bt))/(cosh(al)+cos(bt));
end

function J=lprs2ord2(k,xi,w)
%
% Calculation of a point of the LPRS
% for Transfer Function G(s)=k*s/(s*s+2*xi*s+1),
% w - frequency
% xi < 1
%
if w==0
J=0-j*0;
else
al=pi*xi/w;
sq=sqrt(1-xi*xi);
bt=pi*sq/w;
gm=al/bt;
b=al*cos(bt)*sinh(al)+bt*sin(bt)*cosh(al);
c=al*sin(bt)*cosh(al)-bt*cos(bt)*sinh(al);
denom=sin(bt)^2+sinh(al)^2;
J=0.5*k*(-pi/w*sinh(al)*cos(bt)/denom+xi*(b+gm*c)/denom)...
-j*0.25*k*pi/sq*sin(bt)/(cosh(al)+cos(bt));
end

function J=lprs2ord3(k,w)
%
% Calculation of a point of the LPRS
% for Transfer Function G(s)=k*s/(s+1)^2,
% w - frequency
%
if w==0
J=0-j*0;
else
al=pi/w;
chal=cosh(al);
shal=sinh(al);
J=k*(0.5*al*(-shal+al*chal)/shal/shal-j*0.25*pi*al/(1+chal));
end

function J=lprs2ord4(k,xi,w)
%
% Calculation of a point of the LPRS
% for Transfer Function G(s)=k*s/(s*s+2*xi*s+1),
% w - frequency
% xi > 1
%
```

```
if w==0
J=0-j*0;
else
sq=sqrt(xi*xi-1);
k1=-0.5/sq;
k2=-k1;
t1=xi+sq;
t2=xi-sq;
J=k*(lprs1ord(k1,t1,w)+lprs1ord(k2,t2,w));
end

function J=lprsfopdt(k,t,tau,w)
%
% Calculation of a point of the LPRS
% for transfer function G(s)=k*exp(-tau*s)/(t*s+1),
% 'w' - current frequency
%
if w==0
J=k*(0.5-j*pi/4);
else
al=pi/t/w;
gm=tau/t;
expal=exp(-al);
expgm=exp(gm);
J=0.5*k*(1-al*expgm*csch(al)+j*0.5*pi*(2*expal*expgm/(1+expal)-1));
end
```

Example of the LPRS computing for FOPDT dynamics.

```
script
%
% Calculation of the LPRS of FOPDT dynamics
% with ransfer function
% G(s)=k*exp(-tau*s)/(t*s+1)
% and plotting the locus
%
clear
clc
gain=1.; % gain
tconst=1.; % time constant
tdead=1.; % dead time
ommin=0.0001; % minimum frequency (in this code it is slightly higher
% than 0 to enable the use of the logarithmic scale)
ommax=1000.; % Maximum frequency
% The following code is used to generate logarithmic distribution
% of frequency points
```

```
nom=100; % number of frequency points
lmin=log10(ommin);
lmax=log10(ommax);
delta=(lmax-lmin)/(nom-1);
lom=lmin-delta;
for iom=1:nom
lom=lom+delta;
om=10^lom;
locus(iom)=lprsfopdt(gain,tconst,tdead,om);
end
plot(locus)
grid
axis('equal')
```

The LPRS computing for Example 4.6
```
script
clc
clear
% ————————————————
% Calculation of the LPRS of the
% actuator + plant + sliding surface
% for Example 4.6
% ————————————————
% "w_ex_4_6" is the name of the external
% function (complex) that returns the value of the
% transfer function at frequency "om";
% "ommin" and "ommax" are the low and the high frequencies
% of the analyzed frequency range;
% "nom" is the number of points of this frequency range;
pr=0.; % "pr" is not used in this example;
% however, is can be used for sending parameters to the function
% that calculates the transfer function
ommin=20.;
ommax=10000.;
nom=200;
lmin=log10(ommin); % the next 4 lines are used to make a logarithmic
lmax=log10(ommax); % distribution of frequency points in the range
delta=(lmax-lmin)/(nom-1);
lom=lmin-delta;
name='w_ex_4_6';
for iom=1:nom
lom=lom+delta;
om=10.^lom;
omega(iom)=om;
lprs(iom)=lprsser200(om,name,pr);
```

```
end
plot(lprs)
grid
axis('equal')

function w=w_ex_4_6(om,pr)
s=0+j*om;
wact=1./(0.0001*s*s+2.*0.5*0.01*s+1.);
wplant=1./(s*s+s+1);
w=4.*(s+1.)*wact*wplant/(1.-wact*wplant);
```

References

[1] Anosov D.V. (1959) On stability of equilibrium points of relay systems, Automation and Remote Control, No. 2, 135–149.

[2] Astrom K.J. (1995) Oscillations in Systems with Relay Feedback. The IMA Volumes in Mathematics and its Applications: Adaptive Control, Filtering and Signal Processing, 74: 1–25.

[3] Astrom K.J., Lee T.H., Tan K.K., Johansson K.H. (1995) Recent advances in relay feedback methods - a survey. Systems, Man and Cybernetics, 1995. IEEE International Conference on Intelligent Systems for the 21st Century, 3: 2616–2621.

[4] Astrom K.J., Johanson K.H., Rantzer A. (1999) Fast switches in relay feedback systems. Automatica, 35: 539–552.

[5] Astrom K.J., Hagglund T. (1984) Automatic tuning of simple regulators with specifications on phase and amplitude margins. Automatica, 20: 645–651.

[6] Astrom K.J., Wittenmark B. (1990) Computer-Controlled Systems - Theory and Design. Englewood Cliffs, NJ: Prentice-Hall.

[7] Astrom K.J., Hagglund T. (1995) PID Controllers: Theory, Design and Tuning, second ed. Research Triangle Park, NC: Instrument Society America.

[8] Atherton D.P. (1975). Nonlinear Control Engineering - Describing Function Analysis and Design, Workingham, Berks, UK: Van Nostrand Company Limited.

[9] Atherton D.P. (1993) Analysis and design of relay control systems. In: Linkens D.A. (ed.) CAD for Control Systems. New York: Marcel Dekker.

[10] Barbot J., Djemai M., Boukhobza T. (2002). Sliding mode observers. In: Perruquetti W., Barbot J. (Eds.) Sliding Mode Control in Engineering. New York: Marcel Dekker, 103–130.

[11] Bartolini G., Ferrara A., Usai E. (1998) Chattering avoidance by second-order sliding mode control. IEEE Trans. on Automatic Control, 43 (2), 241–246.

[12] Bartolini G., Ferrara A., Pisano A., Usai E. (2001) On the convergence properties of a 2-sliding control algorithm for nonlinear uncertain systems. Int. J. Control, 74: 718-731.

[13] Bartolini G., Pisano A., Punta E., Usai E. (2003) A survey of applications of second order sliding mode control to mechanical systems. Int. J. Control, 76 (9/10), 875–892.

[14] Bartolini G., Ferrara A., Levant A., Usai E. (1999). On second order sliding mode controllers. In: Young K.D., Ozguner U. (eds.) Variable Structure Systems, Sliding Mode and Nonlinear Control. Berlin: Springer-Verlag, Lecture Notes in Control and Information Sciences, 247: 329–350.

[15] Bartolini G., Ferrara A., Usai E. (1997) Output tracking control of uncertain nonlinear second-order systems. Automatica, 33 (12), 2203–2212.

[16] Di Bernardo M., Johansson K.H., Vasca F. (2001) Self oscillations in relay feedback systems: symmetry and bifurcations, Int. Journal of Bifurcations and Chaos, 11 (4), 1121–1140.

[17] Bilharz H. (1942) Uber eine gesteuerte eindimensionale Bewegung, Zeitschrift fur Angewandte Mathematik und Mechanik, 22 (4), 206–215.

[18] Boiko I. (2003). Frequency-domain analysis of fast and slow motions in sliding modes. Asian Journal of Control, 5 (4), 445–453.

[19] Boiko I. (2004). Analysis of modes of oscillations in a relay feedback system, Proc. 2004 American Control Conference, Boston, MA, USA, 1253–1258.

[20] Boiko I. (2005) Oscillations and transfer properties of relay servo systems - the locus of a perturbed relay system approach, Automatica, 41: 677–683.

[21] Boiko I. (2005). Method and apparatus for tuning a PID controller. US Patent No. 7,035,695.

[22] Boiko I. (2005). Analysis of sliding modes in the frequency domain. Int. J. Control, 78 (13), 969–981.

[23] Boiko I., Fridman L., Castellanos M.I. (2004) Analysis of second order sliding mode algorithms in the frequency domain. IEEE Trans. on Automatic Control, 49 (6), 946–950.

[24] Boiko I., Fridman L. (2005) Analysis of chattering in continuous sliding-mode controllers. IEEE Trans. on Automatic Control, 50 (9), 1442–1446.

[25] Boiko I., Fridman L., Iriarte R., Pisano A., Usai E. (2006) Parameter tuning of second-order sliding mode controllers for linear plants with dynamic actuators. Automatica, 42 (5), 833–839.

[26] Boiko I. (2007) Analysis of closed-loop performance and frequency-domain design of compensating filters for sliding mode control systems. IEEE Trans. on Automatic Control, 52 (10), 1882–1891.

[27] Bondarev A.G., Bondarev S.A., Kostyleva N.E., Utkin V.I. (1985) Sliding modes in systems with asymptotic state observers, Automation and Remote Control, 46, 679–684.

[28] Bromberg P.V. (1953) Stability and self-excited oscillations of impulse control systems, Oborongiz, (in Russian).

[29] Buonomo A., DiBello C. (1996) Asymptotic formulas in nearly sinusoidal nonlinear oscillators. IEEE Trans. Circuits Syst., 43 (8), 953–963.

[30] Burlington R.S. (1973) Handbook of Mathematical Tables and Formulas, New York: McGraw-Hill.

[31] Burton J.A., Zinober A.S.I. (1986) Continuous approximation of variable structure control. Int. J. Syst. Sci., 17, 875–885.

[32] Chua L.O., Tang Y.S. (1982) Nonlinear oscillation via Volterra series. IEEE Trans. Circuits Syst., CAS-29 (2), 150–168.

[33] Chung S.C.-Y., Lin C.-L. (1999) A Transformed lure problem for sliding mode control and chattering reduction. IEEE Trans. on Automatic Control, 44 (3), 563–568.

[34] Chung J.K.-C., Atherton D.P. (1966) The determination of periodic modes in relay systems using the state space approach. Int. J. Control, 4: 105–126.

[35] Coughanowr D.R., Koppel L.B. (1965) Process systems analysis and control, New York: McGraw-Hill.

[36] Dutilh J. (1950) Theorie des servomecanismes arelais. L'Onde electrique, 30: 438–445.

[37] Dwight H.B. (1961) Tables of Integrals and Other Mathematical Data, New York: Macmillan Co.

[38] Edwards C., Spurgeon S. (1998) Sliding mode control: theory and application, London: Taylor & Francis.

[39] Edwards C., Spurgeon. S., Tan C.P. (2002). On development and application of sliding mode observers. In J. Xu and Y. Yu (Eds.), Variable Structure Systems: Towards XXIst Century. Berlin, Germany: Springer-Verlag, 253–282.

[40] Emelyanov S.V., Korovin S.K., A. Levant (1993). Higher order sliding modes in control systems. Differential Equations, 29 (11), 1627–1647.

[41] Fely O., Fitzgerald, D. (1996) Bandpass Sigma-Delta Modulation: An Analysis from the Perspective of Nonlinear Dynamics, Piscataway, NJ: IEEE, 146–149.

[42] Floquet T., Barbot J.-P., Perruquetti W. (2003) Higher-order sliding mode stabilization for a class of nonholonomic perturbed systems. Automatica, 39: 1077–1083.

[43] Flugge-Lotz I. (1953) Discontinuous Automatic Control, Princeton University Press.

[44] Fridman E., Seuret A., Richard J.-P. (2004) Robust sampled-data stabilization of linear systems: an input delay approach, Automatica, 40 (8), 1441–1446.

[45] Fridman L. (1999) The Problem of Chattering: an Averaging Approach. Variable Structure Systems, Sliding Mode and Nonlinear Control. Berlin: Springer-Verlag, 363–386.

[46] Fridman L. (2001). An averaging approach to chattering. IEEE Trans. on Automatic Control, 46 (8), 1260–1264.

[47] Fridman L. (2002). Singularly perturbed analysis of chattering in relay control systems. IEEE Trans. on Automatic Control, 47 (12), 2079–2084.

[48] Fridman L. (2003). Chattering analysis in sliding mode systems with inertial sensors. Int. J. Control, 76 (9/10), 906–912.

[49] Furuta K. (1990). Sliding mode control of a discrete system. Systems and Control Letters. 14: 145–152.

[50] Gelb A., Vander Velde W.E. (1968) Multiple-Input Describing Functions and Nonlinear System Design, New York: McGraw-Hill.

[51] Goldfarb L.S. (1948) On the theory of vibrational regulators. Automation and Remote Control, 9 (6), 413–431 (in Russian).

[52] Goldfarb L.C. (1947) On some nonlinear phenomena in regulatory systems. Automation and Remote Control, 8 (5), (translated from Russian by the National Bureau of Standards, Washington, D.C., Report 1691, 29 May 1952).

[53] Goncalves J.M., Megretski A., Dahleh M.A. (2001). Global stability of relay feedback systems. IEEE Trans. on Automatic Control, 46 (4), 550–562.

[54] Hamel B. (1949). Contribution a l'etude mathematique des systemes de reglage par tout-ou-rien, C.E.M.V., (17), Service Technique Aeronautique.

[55] Hazen H.L. (1934) Theory of servomechanisms. Journal of Franclin Institute, 218: 279–330.

[56] Hsu J.C., Meyer A.U. (1968) Modern Control Principles and Applications, New York: McGraw Hill.

[57] Jury E.I. (1960) A note on the steady-state response of a linear time-invariant system to general periodic input. Proc. IRE, 48 (5), 942–944.

[58] Kalb R.M., Bennett W.R. (1935) Ferromagnetic distortion of a two-frequency wave. Bell System Technical Journal, 14: 322–359.

[59] Kaya I., Atherton D.P. (1999) A PI-PD Controller Design for Integrating Processes. In: Proc. 1999 American Control Conference, San Diego, CA, USA, 258–262.

[60] Kaya I., Atherton D.P. (1998) An improved parameter estimation method using limit cycle data. UKACC Internat. Conf. on Control, IEE, 682–687.

[61] Kaya I., Atherton D.P. (2001) Parameter estimation from relay autotuning with asymmetric limit cycle data. J. Process Control, (11), 429–439.

[62] Khalil H.K. (1996) Nonlinear Systems. Englewood Cliffs, NJ: Prentice Hall.

[63] Kohenburger R.J. (1950) A frequency response method for analyzing and synthesizing contactor servomechanisms. Trans. AIEE, 69: 270–283.

[64] Krylov N.M., Bogolubov N.N. (1937) Introduction to Nonlinear Dynamics, Kiev, Ac. Sc. of Ukrainian SSR.

[65] Kwatny H.G., Young, K.D. (1981) The variable structure servomechanism. Systems and Control Letters, 1 (3), 184–191.

[66] Levant A. (Levantovsky, L.V.) (1993) Sliding Order and Sliding Accuracy in Sliding Mode Control. Int. J. of Control, 58 (6), 1247–1263.

[67] Levant A. (2000). Higher order sliding: differentiation and black-box control. Proceedings of the 39th IEEE Conference on Decision and Control, 2: 1703–1708.

[68] Levant A. (2003) Higher order sliding modes, differentiation and output-feedback control. Int. J. Control, 76 (9/10), 924–941.

[69] Luenberger D.G. (1966). Observers for multivariable systems. IEEE Trans. on Automatic Control, 11 (4) 190–197.

[70] Luyben et al. (1987) Derivation of transfer functions for highly nonlinear distillation columns. Ind. Eng. Chem. Res. 26: 2490–2495.

[71] MacColl L.A. (1945). Fundamental Theory of Servomechanisms, New York: D. Van Nostrand Co.

[72] McDonald D. (1950) Nonlinear techniques for improving servo performance. Proc. National Electronics Conference, Chicago, 6: 400–421.

[73] Majhi S., Atherton D.P. (1999) Autotuning and controller design for processes with small time delays. IEE Proc. – Control Theory Appl, 146 (5), 415–425.

[74] Majhi S., Sahmbi J.S., Atherton D.P. (2001) Relay feedback and wavelet based estimation of plant model parameters. Proceedings of the 40th IEEE Conference on Decision and Control, Florida, USA, 3326–3331.

[75] Majhi S. (2007) Relay-based identification of a class of non-minimum phase SISO processes. IEEE Trans. on Automatic Control, 52 (1), 134–139.

[76] Mandelbrot B.B. (1983) The Fractal Geometry of Nature, New York: W.H. Freeman.

[77] Miloslavljevic C. (2004). Discrete-time VSS. In: Sabanovic A., Fridman L., Spurgeon S. (Eds.), Variable Structure Systems: from Principles to Implementation, IEE, 99–128.

[78] Mitrovski C.D., Kocarev L.M. (2001). Periodic trajectories in piecewise-linear maps. IEEE Trans. on Circuits and Systems, 48 (10), 1244–1246.

[79] Neimark Y.I. (1973). Point Mapping Method in the Nonlinear Oscillations Theory, Moscow: Nauka.

[80] Orlov Y., Aguilar L., Cadiou J.C. (2003) Switched chattering control vs. backlash/friction phenomena in electrical servo-motors. Int. J. Control, 76 (9/10), 959–967.
[81] Pospelov G.S. (1969). Relay Control Systems. In: Automatic Control Theory, Vol. 3, Part II, Moscow: Mashinostroyenie, (in Russian).
[82] Poznyak A.S. (2004) Deterministic output noise effects in sliding mode observation. In: Sabanovic A., Fridman L., Spurgeon S. (Eds.), Variable Structure Systems: from Principles to Implementation, IEE, 45–79.
[83] Qing-Guo W., Chang-Chieh H., Qiang B. (1999). A technique for frequency response identification from relay feedback. IEEE Trans. Control Systems Technology, 7 (1), 122–128.
[84] Rudnev S.A., Faldin N.V. (1998). Linearization of relay servo system with respect to control signal. Russian Ac. Sc. Transactions on Control and Control Systems, (2), 36-43, (in Russian).
[85] Shtessel Y.B., Krupp D.R., Shkolnikov I.A. (2000) 2-Sliding Mode Control for Nonlinear Plants with Parametric and Dynamic Uncertainties. 2000 Conf. AIAA, 1–9.
[86] Shtessel Y.B., Shkolnikov I.A., Brown. M.D.J. (2003) An asymptotic second-order smooth sliding mode control. Asian Journal of Control, 5 (4), 498–5043.
[87] Sidorov I.M., Korotaeva I.P. (1975) Biharmonic oscillations in piecewise-linear systems. International Applied Mechanics, 11 (7), 710–713.
[88] Sira-Ramires H. (1988) Sliding Regimes on Slow Manifolds of Systems with Fast Actuators. International J. of System Science, (37), 875–887.
[89] Slotine J.J.E. (1984) Sliding controller design for nonlinear systems. Int. J. Control, 40: 421–434.
[90] Slotine J.J., Li W. (1991) Applied Nonlinear Control, Prentice Hall. Englewood Cliffs, NJ.
[91] Somerville M.J., Atherton D.P. (1958) Multi-gain representation for a single-valued nonlinearity with several inputs, and the evaluation of their equivalent gains by a cursor method. Proc. lEE, 105C: 537–549.
[92] Su W.C., Drakunov S.V., Ozguner U., Young K.D. (1993) Sliding mode with chattering reduction in sampled data systems. Proceedings of the 32nd IEEE Conference on Decision and Control, San Antonio, Texas, 2452–2457.
[93] Starikova, M.V., Analysis of control systems having logical control devices, Moscow: Mashinostroenie (1978).
[94] Tsypkin Ya.Z. (1984). Relay Control Systems, Cambridge, England: Cambridge University Press.
[95] Tustin A. (1947) The effects of backlash and of speed dependent friction on the stability of closed-cycle control systems. Journal IEE, Pt. IIa, 94: 143–151.
[96] Urabe M. (1965) Galerkin's procedure for nonlinear periodic systems. Arch. Rational Mech. Anal., 20: 120–152.
[97] Utkin V. (1992). Sliding Modes in Control and Optimization, Berlin: Springer-Verlag.
[98] Utkin V.I., Guldner J., Shi J. (1999). Sliding Mode Control in Electromechanical Systems. London: Taylor and Francis.
[99] Utkin V.I. (1983) Sliding mode control: present and future. Automation and Remote Control, 44, 1105–1120.
[100] Varigonda S., Georgiou T.T. (2001). Dynamics of relay relaxation oscillators, IEEE Trans. on Automatic Control, 46 (1), 65–77.

[101] Walcott B.L., Zak H. (1988) Combined observer-controller synthesis for uncertain dynamical systems with applications. IEEE Trans. on Systems, Man and Cybernetics, 18: 88–104.
[102] Wambacq P., Sansen W. (1998) Distortion Analysis of Analog Integrated Circuits. Norwell, MA: Kluwer.
[103] Wang Q.-G., Lee T.H., Lin C. (2003) Relay Feedback: Analysis, Identification and Control. London: Springer.
[104] Weiss H.K. (1946) Analysis of relay servomechanisms. Journal of the Aeronautical Sciences, 13: 364–376.
[105] Young K.D., Utkin V.I., Ozguner U. (1999). A control engineer's guide to sliding mode control. IEEE Trans. Control Systems Technology, 7: 328-342.
[106] Yu C.-C. (1998) Use of Saturation Relay Feedback in PID Controller Tuning, US Patent No. 5742503.
[107] Yu C.-C. (1999) Automatic Tuning of PID Controllers: Relay Feedback Approach. New York: Springer-Verlag.
[108] Zhiltsov K.K. (1974) Approximate Methods for Variable Structure Systems Analysis, Moscow: Energia, (in Russian).
[109] Ziegler J.G., Nichols N.B. (1942) Optimum settings for automatic controllers. Trans. Amer. Soc. Mech. Eng., 64: 759–768.

Index

actuator 68, 77
Additivity property 43
additivity property 23
analogue differentiator 153
asymmetric periodic motion 14
asymmetric relay feedback test 140
asymptotic second-order SM 82, 159
autonomous mode 11, 12
autotune identification 143
autotuning 139, 151

band-pass filters 130
bandwidth 128
bias function 16, 112
boundary layer 103
boundary points 36

cascade compensation 130
chatter smoothing 17
chattering 11, 42, 68, 73, 77, 82, 87, 95, 103, 154, 171
chattering problem 4
closed-loop performance 104, 129, 155
compensating filters 128
compensator 153

describing function 4, 6, 9, 12, 52
discontinuous control 67
Discontinuous control systems 3
discontinuous nonlinearities 3
distributed control systems 139
disturbance attenuation 4, 63, 95

electro-pneumatic servomechanism 127
equivalent control 69

equivalent delay 173
equivalent gain 16, 20, 40, 57, 76, 85, 111, 117, 155
equivalent relay system 73
equivalent time delay 171
execution period 169, 173
existence of a periodic motion 21
external signal propagation 4

fast motions 69
fast signal propagation 57
filtering hypothesis 5
first-order dynamics 33
first-order plus dead time dynamics 38, 141
fixed point 7, 10
Fourier series 8
fractal dynamics 88, 92
frequency domain characteristic 23
frequency of the oscillations 20
frequency-domain approach 6

generalized sub-optimal algorithm 108

Hammerstein model 49
harmonic balance 9, 19, 106
high-frequency segment of LPRS 81
hysteresis 12
hysteresis relay 16
hysteretic relay 13, 58, 109

ideal closed-loop performance 85
ideal sliding mode 42
ideal SM 68, 77
identification 140, 145

Index

infinitesimally small constant input 17
initial perturbation 26
input-output analysis 12, 63
input-output problem 4, 5, 68
integrating linear part 26

Jacobian matrix 26, 30

limit cycle 78, 104
linear part 7, 8, 79, 128
linearized model of the relay servo system 63
locus of a perturbed relay system 20
low-pass filtering 121, 154, 159
low-pass filtering property 10
LPRS 20, 79, 93, 140
LPRS extended definition 46

negative reciprocal of the DF 13
non-ideal closed-loop performance 77
non-ideal disturbance rejection 77
non-ideal tracking 77
non-integrating linear part 29
non-reduced-order model 16, 85
nonlinear part 7, 8
nonlinear plant 43, 48, 135
nonlinear systems 3

observation error 174
on-off control 3
open-loop gain 4
open-loop LPRS computing 46
operational amplifier 154
orbital stability 26, 30
oscillations 77
overshoot 147

parallel compensation 131
parasitic dynamics 70, 79, 85, 88, 103, 154, 168
partial fractions 23, 37
periodic motions 4
periodic problem 5, 6
periodic signal mapping 7, 48
periodic solution 7, 13, 36
periodic solution of finite frequency 36
Poincaré map 4, 8, 21
Poincaré return map 21, 39, 74
principal dynamics 78, 85, 154, 167

real SM 68, 77
reduced order dynamics 72

reduced-order model 85
reduced order system 68
relative degree 81, 86, 88, 103
relay feedback system 3, 5
Relay feedback test 139
relay servo system 11, 171
Relay servomechanisms 125

sampling rate 175
second-order dynamics 35
second-order SM 103
self-excited oscillations 139, 156
self-similarity 89
sensor 68
servo problem 63
servo systems 10
servomechanism performance 127
set point tracking 4
settling time 149
singular perturbation method 6
sliding mode 3, 67
sliding mode control 4
sliding surface 67, 70
sliding variable 79
slow motions 69
slow signals propagation 17
SM compensator 158
SM differentiator 153, 156
SM observer 167
spectral domain 9
stability of an equilibrium point 26
stability of limit cycles 4
stabilization systems 10
state estimates 169
static gain 42, 141
sub-optimal algorithm 104
subspace controls 71
switching imperfections 68
switching surface 67

time delay 31, 38
transfer function 24, 29
Tsypkin locus 5, 52
Tuning criterion 147

uncompensated servomechanism 130
unequally spaced switching 28

variable structure control 67
variable structure systems 4

Wiener model 49

Printed in the United States of America